Contemporary Dielectric Materials

Edited by R. Saravanan

This book deals with experimental results of the physical characterization of several important, dielectric materials of great current interest. The experimental tools used for the analysis of these materials include X-ray diffraction, dielectric measurements, magnetic measurements using a vibrating sample magnetometer, optical measurements using a UV-Visible spectrometer etc.

The materials studied and reported in this book are as follows; the impedance analysis of nanocrystalline NiO prepared using the combustion method; PL (photoluminescence, IR (Infra-red, Raman, and X-ray characterization of GaO powders prepared using the chemical method; X-ray, SEM (Scanning Electron Microscopy), VSM (Vibrating Sample Magnetometer), UV-Vis (UltraViolet-Visible) characterization of the multiferroic material $Ga_{2-x}Fe_xO_3$ prepared using the SSR (Solid State Reaction) method; XRD and optical studies on sol-gel prepared samarium and manganese substituted calcium hydroxyapatite; defect studies and positron annihilation studies on ZnO nano particles prepared using the sol-gel and combustion methods; Bonding in $La_{0.9}Zn_{0.1}FeO_3$ multiferroic material prepared using the chemical method; effect of temperature on the magnetic phase transition in $Co_{0.5}Zn_{0.5}Fe_2O_4$ prepared using the mechanical alloying method; effect of sintering temperature on the micro structure and optical properties of ZnO ceramics.

Contemporary Dielectric Materials

Edited by

Dr. R. Saravanan, M.Sc., M.Phil., Ph.D.
Associate Professor & Head
Research Centre and PG Department of Physics
The Madura College (Autonomous)
Madurai - 625 011
India

Published by **Materials Research Forum LLC**
Millersville, PA 17551, USA

Published as part of the book series
Materials Research Foundations
Volume 7 (2017)
ISSN 2471-8890 (Print)
ISSN 2471-8904 (Online)

Print ISBN 978-1-945291-12-8
ePDF ISBN 978-1-945291-13-5

This book contains information obtained from authentic and highly regarded sources. Reasonable efforts have been made to publish reliable data and information, but the author and publisher cannot assume responsibility for the validity of all materials or the consequences of their use. The authors and publishers have attempted to trace the copyright holders of all material reproduced in this publication and apologize to copyright holders if permission to publish in this form has not been obtained. If any copyright material has not been acknowledged please write and let us know so we may rectify in any future reprint.

Distributed worldwide by

Materials Research Forum LLC
105 Springdale Lane
Millersville, PA 17551
USA
http://www.mrforum.com

Manufactured in the United State of America
10 9 8 7 6 5 4 3 2 1

Table of Contents

Preface

This book deals with experimental results of the physical characterization of several important, dielectric materials. The experimental tools used for the analysis of these materials include X-ray diffraction, dielectric measurements, magnetic measurements using a vibrating sample magnetometer, optical measurements using a UV-Visible spectrometer etc.

The materials studied and reported in this book are as follows; the impedance analysis of nanocrystalline NiO prepared using the combustion method; PL (photoluminescence, IR (Infra-red, Raman, and X-ray characterization of GaO powders prepared using the chemical method; X-ray, SEM (Scanning Electron Microscopy), VSM (Vibrating Sample Magnetometer), UV-Vis (UltraViolet-Visible) characterization of the multiferroic material $Ga_{2-x}Fe_xO_3$ prepared using the SSR (Solid State Reaction) method; XRD and optical studies on sol-gel prepared samarium and manganese substituted calcium hydroxyapatite; defect studies and positron annihilation studies on ZnO nano particles prepared using the sol-gel and combustion methods; Bonding in $La_{0.9}Zn_{0.1}FeO_3$ multiferroic material prepared using the chemical method; effect of temperature on the magnetic phase transition in $Co_{0.5}Zn_{0.5}Fe_2O_4$ prepared using the mechanical alloying method; effect of sintering temperature on the micro structure and optical properties of ZnO ceramics.

This volume includes 10 chapters dealing with the experimental growth and characterization of the above mentioned samples. NiO nanostructures prepared using different methods, like sol-gel method and combustion method have been experimentally studied and reported in this book. Several dielectric ceramics like $Ga_{1-x}Fe_xO$, ZnO bulk and ZnO nano particles nano TiO_2 and the multi ferroic compounds $La_{1-x}Zn_xFeO_3$ have been studied and reported as well. The different chapters were written by active scientists and researchers from reputed Institutions from India, South Korea, Lithuania, Taiwan, Malaysia, and Indonesia.

Dr. R. Saravanan, Madurai, India

CHAPTER 1

Electrical impedance analysis for nano crystalline NiO prepared by combustion method

N. Nallamuthu*, S. Asath bahadur, V. Siva, A. Shameem

Department of physics, Kalasalingam University, Krishnankoil – 626126, Tamilnadu, India

Email: nnallamuthu@gmail.com*

Abstract

Nanocrystalline NiO material has been synthesized using a tartaric acid based sol-gel combustion method. The obtained XRD peaks for various calcinating temperatures have been analyzed and reported. The XRD particle size is determined by using the Scherrer formula and it is found to be ~27nm at 600 °C. The structural identification of NiO has been done using Fourier transform infrared spectroscopy (FTIR). The spherical shaped agglomerated NiO particles are characterized by using the SEM technique and the atomic percentages are verified by EDX techniques. The electrical conductivities of sintered NiO are evaluated through an impedance analysis and the activation energy is found to be 0.21 eV.

Keywords

Nanocrystalline, XRD, SEM, Impedance, Electrical Conductivity, Cathode Materials

Contents

1. Introduction

Nickel oxide is one of the most interesting transition metal oxides and it has enormous attention due to its usage in different applications such as magnetic devices [1], electro chromic films [2], gas sensors [3,4], dye sensitized photo cathodes [5,6], catalysis [7], etc. Nano crystalline nickel oxide has more enhanced properties than the micro sized particles because of the quantum size effect, surface effect, macroscopic quantum tunneling effect and volume effect [8]. The reduction of the grain size in nano sized NiO material is better for the capacitive enhancement and has proven more advantages for enhancing electrochemical energy storage properties including power and energy density [9, 10]. It has a high ion and electron transport because of the large surface to volume ratio and also the formation of a short diffusion pathway between the nano grains [11,12]. Nano materials are easily adopted in materials science because of their novel properties due to their size and shape. Also, new synthetic approaches are available for getting nano samples and to enhance their growth mechanisms, etc. Thus, several synthesis methods are attempted to prepare the nano particles in different forms such as nano grain, nano rod, nano sheet, etc. Dual fuel based combustion is one of the methods to prepare nano crystalline materials [13]. Citric acid and tartaric acid are used as a fuel and capping agent to make very fine nano sized metal oxide particles. In this study, an attempt has been made to synthesize nano crystalline NiO materials by a citric acid and tartaric acid based combustion process.

2. Material synthesis

Analytical grade nickel nitrate was used as a precursor material and citric acid and tartaric acid are used as fuels of the tartaric acid based sol-gel combustion method. Fig 1. shows the preparation of nano crystalline NiO powder by the combustion method. Stochiometric amount of nickel nitrate was added with distilled water and the solution was marked as label A. The citric acid and tartaric acid were weighed as per 1:1 metal ion ratio and they were dissolved with distilled water. The solution was named as B. Solution B was added drop wise to solution A and the final mixture was stirred continuously under heating resulting in a transparent resin. The sticky resin was dried at 150° C at a heating rate of 1°/min. Then, it was calcined at different temperatures and characterized by various techniques such as XRD, FTIR, SEM-EDX. Electrical conductivity of the sample was analyzed using the experimentally measured impedance parameters. The powder X-ray diffraction (XRD) patterns were obtained using Bruker made X-ray diffractometer (D8 advanced ECO XRD systems with SSD160 1 D Detector) with monochromatic Cu Kα radiation (λ =1.5418A) at diffraction angle between 5° to 120°. Infrared spectroscopic analysis was carried out with a Schimadzu FTIR-8700 Fourier transform infrared spectrometer in the wavelength range 800 to 350 cm^{-1}. For electrical conductivity analysis, the nano crystalline NiO sample was pelletized and sintered at 600° C and the sintered pellets were painted with silver paste and annealed at 300° C. The electrical impedance was measured using Hioki made impedance analyzer in the frequency range 40 Hz to 1MHz.

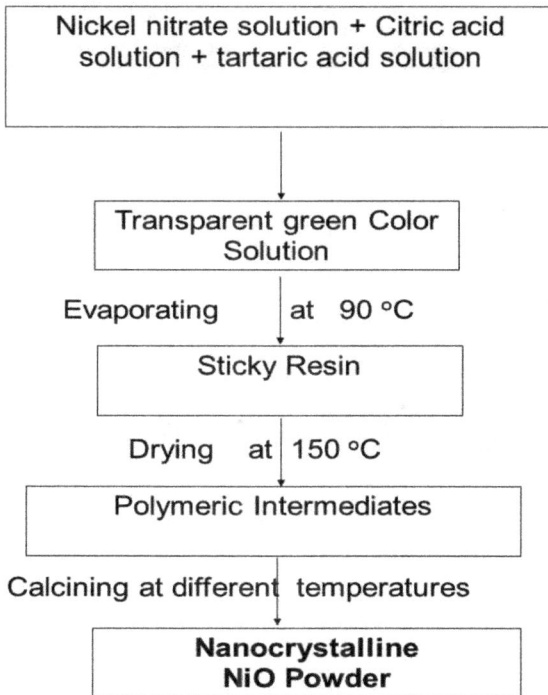

```
┌─────────────────────────────────────────┐
│   Nickel nitrate solution + Citric acid  │
│   solution + tartaric acid solution      │
│                                          │
└─────────────────────────────────────────┘
                      │
                      ▼
        ┌─────────────────────────────┐
        │  Transparent green Color     │
        │  Solution                    │
        └─────────────────────────────┘
  Evaporating          │ at   90 °C
        ┌─────────────────────────────┐
        │       Sticky Resin           │
        └─────────────────────────────┘
       Drying    at │ 150 °C
    ┌─────────────────────────────────┐
    │   Polymeric Intermediates        │
    └─────────────────────────────────┘
  Calcining at different  temperatures
        ┌─────────────────────────────┐
        │      Nanocrystalline         │
        │      NiO Powder              │
        └─────────────────────────────┘
```

Figure. 1 Preparation of nano crystalline NiO powder by the sol-gel combustion method

3.   Results and discussion

3.1   X-ray diffraction studies

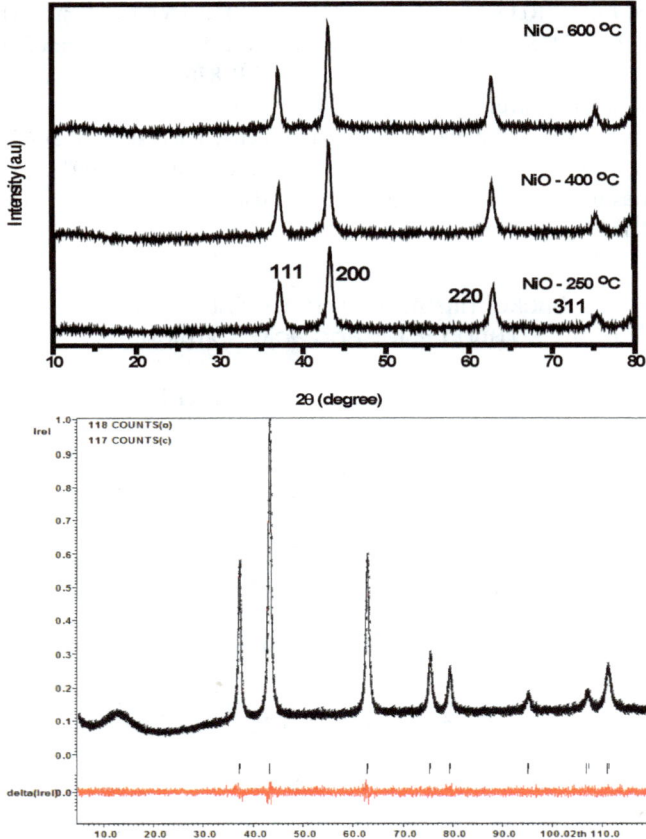

Figure. 2   XRD patterns for NiO sample, calcined at various temperatures and Rietveld refinement profile, calcined at 600° C.

The observed diffraction patterns and Rietveld refined diffraction patterns are shown in fig.2. The observed broad lines signify that the crystallite size is small. Additional X-ray diffraction peaks other than allowed for the system are not observed in the X-ray patterns, this fact indicates the high degree of phase purity in the synthesized NiO materials. The observed X-ray peaks indicate the formation of cubic phase NiO nano particles and this is verified through the JCPDS data (#00-047-1049) [14]. All the observed peaks were indexed as face centered cubic structure of NiO crystalline material. The lattice parameter of NiO is found to be a= 4.2154°. The observed peaks confirm that the precursors and fuels have no role to play in the crystal structure. The crystallite size is also evaluated using the Scherrer formula and it is found to be ~27nm for the sample calcined at 600° C.

3.2 FTIR studies

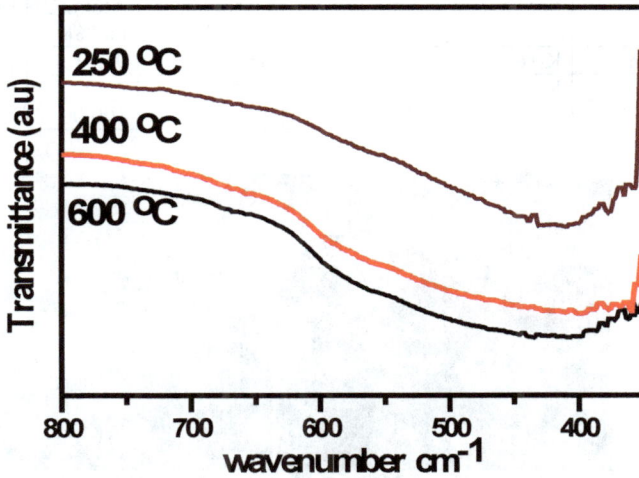

Figure. 3   FTIR Spectra for NiO sample, calcined at various temperatures.

Fig. 3. shows the FTIR spectra for NiO materials, calcined at different temperatures within the frequency range of 800 cm^{-1} to 350 cm^{-1}. The observed broad FTIR peak at 450 cm^{-1}, indicates the formation of a NiO band [15]. At increased temperatures, the broadening of the FTIR peaks also increases, which confirms the growth of NiO particles.

3.3   SEM-EDX analysis

SEM images of the NiO samples, calcined at 250°, 400° and 600° and the STEM image of NiO sample, calcined at 600° are shown in figs.4a-d. The SEM images show spherical shaped partciles, having size ~ 50nm. With increasing sintering temperature, the particle size also increases. STEM image also show a particle size of around ~ 50nm. EDX spectrum confirmed the existence of Ni and O in the NiO sample, calcined at 600° as shown in fig. 5. Table 1 of EDX data shows the individual atomic percentage of NiO sample, which confirms the stoichiometry of the synthesized nano crystalline NiO sample.

Table 1: Atomic weight percentage of elements obtained from EDX spectra.

Element	Series	unn. C [wt.%]	norm. C [wt.%]	Atom. C [at.%]	Error (3 Sigma) [wt.%]
Oxygen	K-series	17.38	85.30	95.51	7.24
Nickel	K-series	3.00	14.70	4.49	0.51
Total		20.37	100.00	100.00	

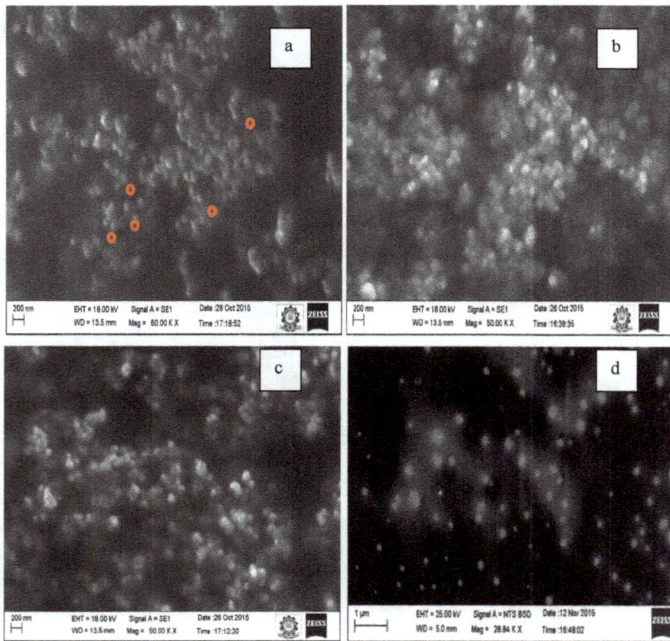

Figure. 4 a,b,c,d   SEM images for NiO samples, calcined at 250 °C, 400 °C and 600 °C and STEM image of the NiO sample at 600 °C.

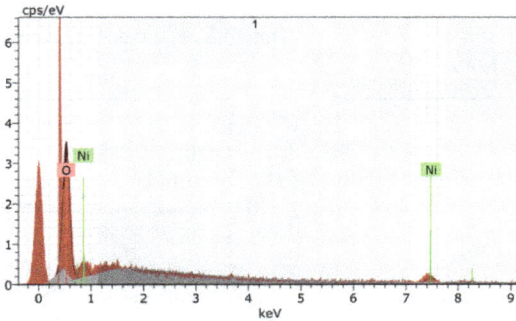

Figure. 5   EDX spectum for NiO samples, calcined at 600 °C.

3.4   Electrical impedance studies

Fig.6 shows the electrical impedance plot for the NiO sample, heat treated at different temperatures. The value of the impedance decreases while the sintering temperature increases. The complex impedance data is fitted using Z- data software.

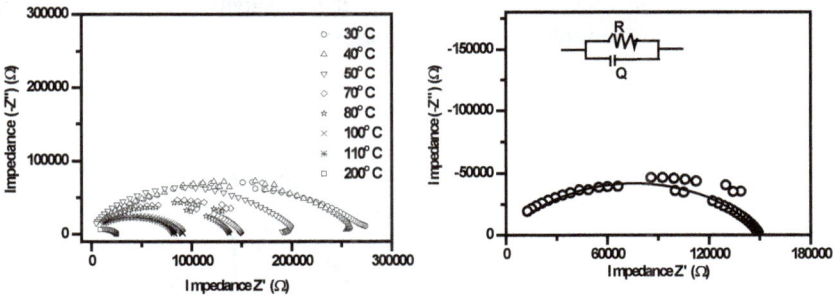

Figure. 6   Electrical impedance plot for NiO samples, sintered at 600 °C and Figure. 7 Electrical impedance plot with Equivalent circuit for NiO samples, heat treated at 70°C.

Fig 7. shows that the fitted electrical impedance plot of the NiO sample, heat treated at 70°. The impedance plot shows a semicircle which is equivalent to parallel combination of a resistance and a capacitance. The bulk impedance is measured by extension of the semicircle on the x-axis [16]. Bulk impedances of the sample are measured for NiO nano materials, heated at various temperatures. The bulk conductivity is also calculated using the relation $\sigma = 1/RA$. Fig.8. shows log (σT) vs 1000/T plot for the NiO sample, sintered at 600°. Using the plot of conductivity vs temperature, the activation energy is evaluated and is found to be 0.213eV.

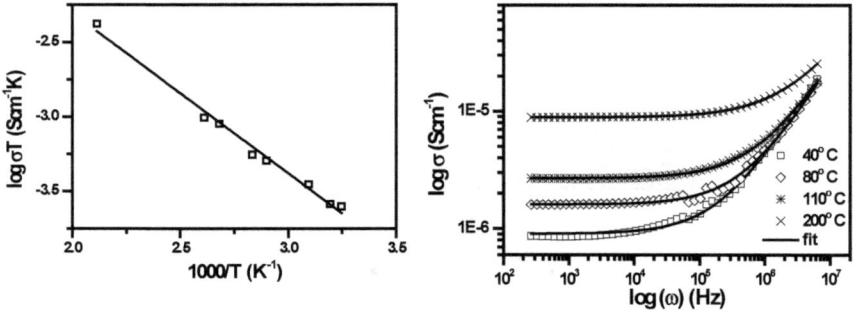

Figure. 8   log(σT) vs 1000/T plot for nano crystalline NiO sample, sintered at 600° C and Figure. 9 log (σ) vs log(ω) plot for nano crystalline NiO sample, sintered at 600° C.

Fig 9. shows the plot of ac conductivity for various frequencies of NiO nano materials. The DC conductivity at lower frequencies increases with rising temperature. Fig.10. shows the plot of frequency dependent ac electrical conductivity. It is seen that the peak frequencies are shifted towards higher frequencies with rising temperature.

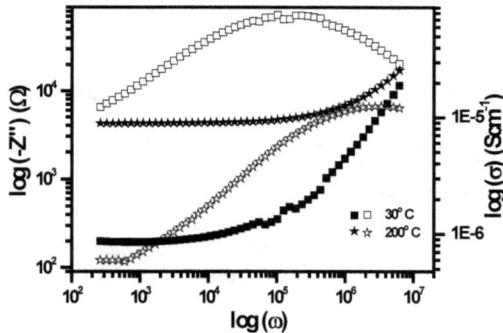

Figure. 10 log σ, log (-Z") vs logω plot for sintered nano crystalline NiO sample, heat treated at difference temperatures.

3.5 Transference number studies

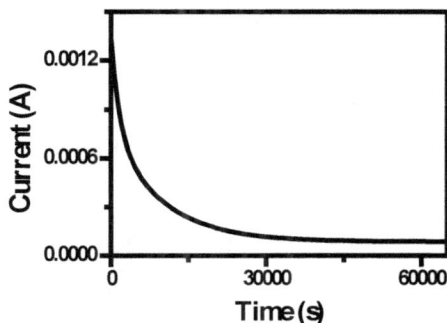

Figure 11. *Polarization current as a function of time for nano crystalline NiO sample at RT.*

The ionic and electronic transference numbers have been evaluated for nano crystalline NiO sample pellet, sintered at $600°\,C$ using the DC polarization technique by applying an voltage of 1V. Fig. 11 shows the polarization current as a function of time for nano crystalline NiO sample. The decreasing polarization current can be attributed to migration of ions due to the applied field and balanced by the diffusion due to the concentration gradient [17]. Thus, the resulting current is only due to the electrons and holes. The ionic and electronic transference numbers are calculated by DC polarization techniques using the relation $t_i = (I_f - I_e)/I_t$ and $t_e = (I_f/I_e)$. The ion transference number is found to be 0.93 and electron transference number 0.07.

4. Conclusions

Nano crystalline NiO material has been synthesized by the tataric acid based combustion method. The recorded XRD pattern confirms that the obtained peaks correspond to NiO with cubical structure. FTIR peaks confirm the NiO band structure. SEM images show agglomerated nano sized spherical particles. EDX spectrum has confirmed the presence of stoichimetric Ni and O elements in the NiO sample. Bulk impedance and conductivities have been analyzed and the activation energy is found to be 0.21eV. The transference numbers have been evaluated through DC polarization techniques. It indicates the possibility of ionic and electronic motion in the NiO samples. These results can be applied to cathode materials in a solid state battery.

References

[1] J. Sort, S. Surinach, J. S. Munoz, M. D. Baro, J. Nogues, G. Chouteau, V. Skumryev and G. C. Hadjipanayis, Improving the energy product of hard magnetic materials physical review b, 65, (2002) 174420J.

[2] M. Chigane and M. Ishikawa, Characterization of electrochromic nickel oxide thin films prepared by anodic deposition, J. Chem. Soc., Faraday Trans., (1992) 88, 2203.
http://dx.doi.org/10.1039/ft9928802203

[3] B. C. Alcock, Z. Li, W. J. Fergus, and L. Wang, New electrochemical sensors for oxygen determination, Solid State Ionics, 39, (1992) 53.
http://dx.doi.org/10.1016/0167-2738(92)90362-s

[4] H. Kumagai, M. Matsumoto, K. Toyoda, and M. Obara, Preparation and characteristics of nickel oxide thin film by controlled growth with sequential surface chemical reactions, J. Mater. Sci. Lett., 15, (1996) 1081
http://dx.doi.org/10.1007/BF00274914

[5] J. He, H. Lindstrom, A. Hagfeldt, S.-E. Lindquist, Dye-sensitized nanostructured p-type nickel oxide film as a photocathode for a solar cell, J. Phys. Chem B 103, (1999) 8940.
http://dx.doi.org/10.1021/jp991681r

[6] Z. Ji, G. Natu, Y. Wu, Cyclometalated ruthenium sensitizers bearing a triphenylamino group for p-type NiO dye-sensitized solar cells, ACS Appl. Mater. Interfaces 5, (2013) 8641.
http://dx.doi.org/10.1021/am402263q

[7] C. Nitin, S. Sunayana, M.K. Sharma, R.K. Chaturvedi, Photocatalytic degradation of safranine O in the presence of nickel oxide, Int. J. Res. Chem. Environ. 1, (2011) 66.

[8] V. Srinivasan, J. Weidner, An electrochemical route for making porous nickel oxide electrochemical capacitors, J. Electrochem. Soc. 144, (1997) L210.
http://dx.doi.org/10.1149/1.1837859

[9] K. Soulantica, L. Erades, M. Sauvan, F. Senocq, A. Maisonnat, B. Chaudret, Synthesis of indium and indium oxide nanoparticles from indium cyclopentadienyl precursor and their application for gas sensing, Adv. Funct. Mater. 13 (2005) pp. 553–557.
http://dx.doi.org/10.1002/adfm.200304291

[10] M. Epifani, E. Comini, J. Arbiol, R. Diaz, N. Sergent, T. Pagnier, P. Siciliano, G. Faglia, J.R. Morante, Chemical synthesis of In2O3 nanocrystals and their application in highly performing ozone-sensing devices, Sens. Actuators B 130, (2008) 483.
http://dx.doi.org/10.1016/j.snb.2007.09.025

[11] C. Xu, J. Tamaki, N. Miura, N. Yamazoe, Grain size effects on gas sensitivity of porous SnO 2-based elements, Sens. Actuators B 3 (1991) 147.
http://dx.doi.org/10.1016/0925-4005(91)80207-Z

[12] N. Yamazoe, New approaches for improving semiconductor gas sensors, Sens. Actuators B 5, (1991) 7.
http://dx.doi.org/10.1016/0925-4005(91)80213-4

[13] A. E. Danks, S. R. Hall and Z. Schnepp, The evolution of 'sol–gel'chemistry as a technique for materials synthesis, Mater. Horizon 3, (2016) 91.
http://dx.doi.org/10.1039/C5MH00260E

[14] S. Saravanakumar, R. Saravanan, S. Sasikumar, Effect of sintering temperature on the magnetic properties and charge density distribution of nano-NiO, Chemical papers 68, (2014) 788.
http://dx.doi.org/10.2478/s11696-013-0519-1

[15] S. Mohseni Meybodi, S. A. Hosseini, M. Rezaee, S. K. Sadrnezhaad, D. Mohammadyani, Synthesis of wide band gap nanocrystalline NiO powder via a sonochemical method, ultrasonics sonochemistry 19, (2012) 841.

[16] N. Nallamuthu, I. Prakash, M. Venkateswarlu, S. Balasubramanyam, N. Satyanarayana, Sol–gel synthesis and characterization of Li 2 O–As 2 O 5–SiO 2 glassy system, 9, (2008) 15.

[17] M. Muthuvinayagam, C. Gopinathan, Characterization of proton conducting polymer blend electrolytes based on PVdF-PVA, Polymer (2015) 68, 122.
http://dx.doi.org/10.1016/j.polymer.2015.05.008

CHAPTER 2

Synthesis and structural characterization of gallium oxide powders

R. Yuvakkumar[1,2*], D. Sidharth[1], G. Ravi[1], S.I. Hong[2*]

[1]Nanomaterials Laboratory, Department of Physics, Alagappa University, Karaikudi - 630 004, Tamil Nadu, India

[2]Department of Nanomaterials Engineering, Chungnam National University, Daejeon, 305-764, South Korea

Email: yuvakkumar@gmail.com, sihong@cnu.ac.kr

Abstract

A potential green synthesis of β-Ga_2O_3 powders has been achieved employing the gallium-ellagate complex formation method. A possible mechanism has been proposed to understand the formation of gallium-ellagate complex formation. The effect of reaction time, incubation and calcination temperature on the product output was explored. The structure and crystalline phase of β-Ga_2O_3 powders were revealed from the X-ray diffraction and from photo luminescence, Raman and IR spectroscopy. The structural characterization revealed the monoclinic phase of Ga_2O_3 with preferential growth along (111).

Keywords

Gallium oxide, Green Synthesis, Structural properties, Gallium-ellagate, Calcination

Contents

1.   Introduction

Gallium oxide (Ga_2O_3) has been attracting considerable attention in optoelectronics, gas sensor, luminescent, nonvolatile memory, and heterogenous catalyst and oxide electronics applications [1-3]. Moreover, it has also been used in military and telecommunication services and also in automatic, bio metric and proximity access control systems [4-6]. Recently, Pallister and his co-workers investigated self-seeding gallium oxide nanowire growth employing pulsed chemical vapor deposition [7]. Szwejkowski and his co-workers studied the effects of material size in the thermal conductivity of gallium oxide (β-Ga_2O_3) films grown via open-atmosphere annealing of gallium nitride [8]. In their work, they studied the thermal conductivity of beta-phase gallium oxide (β-Ga_2O_3) thin films as a component of typical gate oxides used in semiconductor devices [8]. Lawrenz and his co-workers investigated the morphology, mechanical stability and protective properties of ultrathin gallium oxide coatings [9]. They explored that the ultrathin gallium oxide layers could be a promising candidate for protective layers in flexible organic (opto-) electronics and photovoltaics due to permeation barrier functionalities in conjunction with high optical transparency [9]. In addition, Golubev and his co-workers investigated the diffusion-driven and size-dependent phase changes of gallium oxide nanocrystals in a glassy host [10]. Cai and his co-workers investigated the capacitive behavior of single gallium oxide nano belt [11]. Xu and his co-workers investigated the aqueous solution-deposited gallium oxide dielectric for low-temperature, low-operating-voltage indium oxide thin-film transistors employing a facile route to green oxide electronics [12].

In the present study, the inorganic complex intermediate green ligation from natural waste resources has been adopted to synthesis gallium oxide powders [13-21]. Among many synthetic methods, metal-ellagate square planar complex method is the simple technique to prepare bio inspired transition metal oxide powders. Therefore, in the present study, we report an effective metal oxide powder synthesis at room temperature incubated 7 days via gallium-ellagate square planar complex method. In the present study, the obtained gallium-ellagate complex has been decomposed into Ga_2O_3 powders by calcination in a static air atmosphere at 350°, 500° and 650° C for 1 h.

2.   Materials and methods

2.1   Synthesis of gallium oxide powders

Gallium (III) nitrate hydrate $[Ga(NO_3)_3.nH_2O]$ and ethanol were purchased from Merck chemicals. Rambutan collected from a Daejeon supermarket in South Korea were manually separated into peels and subsequently placed in a circulating oven at 50° C. Finely dried rambutan peels (3g) were boiled with ethanol and double distilled water mixture (1:2 ratio) for 10 min. 0.1 M gallium nitrate hydrate was prepared in 50 mL and 10 mL rambutan extract was slowly added drop wise into the solution under magnetic stirring at 80° C for 2 h and then incubated at room temperature for 7 days to form gallium-ellagate complex formation. Gallium oxide powders were obtained due to direct decomposition of gallium-ellagate complexes in muffle furnace at 350°, 500° and 650° C for 1 h in a static air atmosphere.

2.2.   Characterization of gallium oxide powders

Phase and crystalline nature of prepared samples were identified by X-ray powder diffraction (X'pert PRO analytical diffractometer) using CuKα as radiation (1.541 A°) source. Imaging spectrograph STR 500 nm focal length laser Micro Raman spectrometer SEKI, Japan with resolution: 1/0.6 cm-1/pixel and Flat Field: 27 mm (W) × 14mm (H) was used. Photoluminescence (PL) spectrum was measured using Varian Cary Eclipse photoluminescence spectrometer with Oxford low temperature LN277K setup. Infrared (IR) spectra were recorded using Fourier transform infrared spectrophotometer using Thermo Nicolet 380 with resolution 0.5 cm-1 and S/N ratio: 2000:1 ppm for 1 minute scan.

3.   Results and discussion

3.1   XRD pattern of gallium oxide powders

The phase and crystalline nature were identified employing X-ray powder diffraction patterns. The X-ray diffraction patterns of the sample were recorded using a computer controlled X'pert PRO analytical diffractometer using Cu-Kα (wavelength = 1.541 A°) radiation as source and operated at 40 KV. The scanning range was between 10 and 80 degrees. The overall structure of the synthesized product was characterized by XRD. The structure and phase variation of Ga_2O_3 powders incubated at different temperature calcinated at 350° C, 500° C and 650° C are shown in Fig. 1(a-c). It can be seen that the whole spectrum can be indexed from a peak position to a monoclinic crystalline Ga_2O_3 phase, which is in good agreement with reported values of β - Ga_2O_3 with lattice constant a=12.23 Å, b=3.04 Å, c= 5.80 Å and β=103.7 (JPCDS: 43-1012).

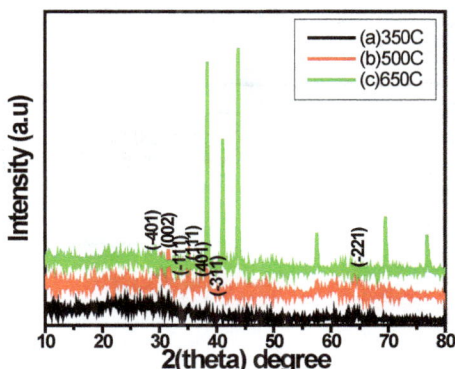

Figure. 1   XRD Pattern: 7 days of reaction product from the mixture of 0.1M gallium nitrate hydrate and 10 ml extract calcinated at (a) 350° C, (b) 500° C and (c) 650° C.

The phase identity of the reaction product (incubated for 7 days) obtained from the mixture of 0.1M gallium nitrate and 10 ml extract and calcinated at 500° C is shown in Fig.1(b). Miller indices were indicated on each diffraction peak. The XRD pattern can be indexed to the monoclinic single crystalline phase of β-Ga_2O_3 with lattice constants a = 4.58 Å, b = 9.80 Å and c = 2.97 Å which is in good agreement with the reported value of the β-Ga_2O_3 (JCPDS: 43-1012). No other peaks of impurities and no other phase of Ga_2O_3 were found within the detection limit of 2θ in the range 10-80°. The most intense peak (111) indicates the growth direction of the sample. The average crystallite size was calculated for the prominent plane (111) using Debye-Scherrer formula and was found to be 90 nm.

3.2   PL spectrum of gallium oxide powders

Fig. 2 exhibits room temperature Photo Luminescence (PL) spectrum of Ga_2O_3 powders incubated at different temperature of 350° C, 500° C and 650° C. The excitation was conducted under 250 nm UV light from a Xe lamp. A broad and strong yellow and orange emission band centered at 575 nm was obtained. The luminescence properties of β- Ga_2O_3 have been extensively studied for several decades and an acceptable model for yellow and orange emission has been put forward [22]. The PL may originate from the recombination of electron on a donor formed by oxygen vacancies and hole on an acceptor formed by gallium vacancies or by gallium-oxygen vacancy pair [23]. It is expected that present preparation method would easily produce considerable quantity of oxygen vacancies and gallium-oxygen vacancy pairs due to high temperature and oxygen – deficient pair.

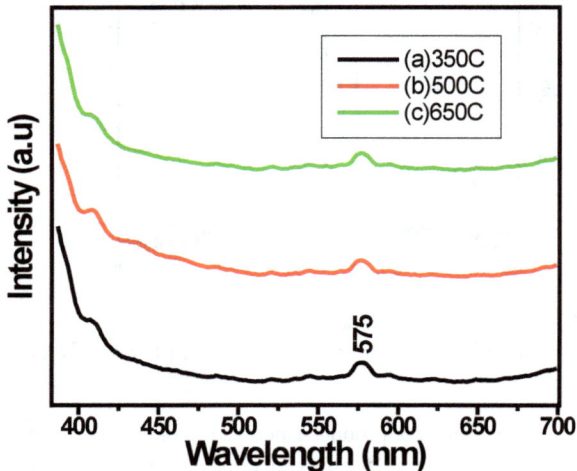

Figure. 2   Room-temperature PL spectra: 7 days of reaction product from the mixture of 0.1M gallium nitrate hydrate and 10 ml extract calcinated at (a) 350° C, (b) 500° C and (c) 650° C

3.3   Raman spectrum of gallium oxide powders

In addition, the structural and electronic properties of Ga_2O_3 powders incubated at different temperature over a period of 350° C, 500° C and 650° C were examined employing Raman spectroscopy. Fig. 3(a-c) show the Raman spectra of Ga_2O_3 powders incubated at various temperatures namely 350° C, 500° C and 650° C. Fig. 3(a-c) reveal the micro-Raman spectra of the product calcinated at (a) 350° C, (b) 500° C and (c) 650° C which confirms the Raman vibration modes of β-Ga_2O_3.

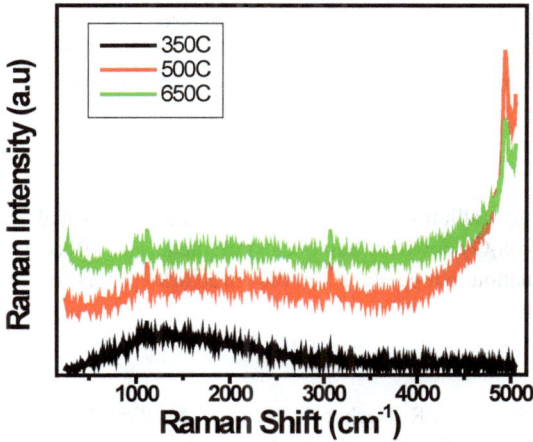

Figure.3   Raman Spectra: 7 days of reaction product from the mixture of 0.1M gallium nitrate hydrate and 10 ml extract calcinated at (a) 350° C, (b) 500° C and (c) 650° C.

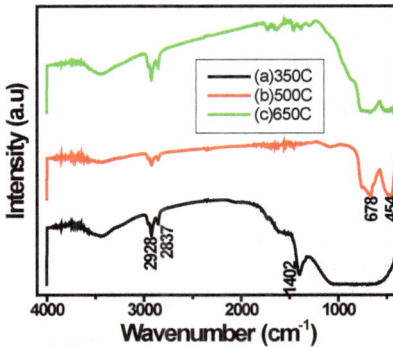

Figure. 4   IR Spectra: 7 days of reaction product from the mixture of 0.1M gallium nitrate hydrate and 10 ml extract calcinated at (a) 350° C, (b) 500° C and (c) 650° C.

3.4   IR spectrum of gallium oxide powders

Moreover, the FTIR spectrum of β-Ga_2O_3 powders was recorded in the range 400 to 4000 cm^{-1} as shown in Fig. 4(a-c) from which the functional groups were identified. The peaks around 2928 cm^{-1} and 2837 cm^{-1} can be assigned to the stretching vibration of the H-O-H group. The weak band at 1402 cm^{-1} represents the bending vibrations of adsorbed molecular water. The strong peak at 678 cm^{-1} represents the Ga-O-Ga bending vibration and the peak at 454 cm^{-1} can be assigned to Ga-O bending vibration.

4. Conclusions

A cost effective facile gallium-ellagate complex formation assisted green synthesis method to prepare β-Ga$_2$O$_3$ powders has been presented. The effect of reaction time, incubation and calcination temperature on the structural properties of the obtained product was investigated.

Reference

[1]   Z. Galazka, K. Irmscher, R. Uecker, R. Bertram, M. Pietsch, A. Kwasniewski, M. Naumann, T. Schulz, R. Schewski, D. Klimm, M. Bickermann, On the bulk β-Ga2O3 single crystals grown by the Czochralski method, J. Cryst. Growth 404 (2014) 184–191.
       http://dx.doi.org/10.1016/j.jcrysgro.2014.07.021

[2]   Y. Tomm, P. Reiche, D. Klimm, T. Fukuda, Czochralski grown Ga2O3 crystals, J. Cryst. Growth 220 (2000) 510–514.
       http://dx.doi.org/10.1016/S0022-0248(00)00851-4

[3]   E.G. Víllora, K. Shimamura, Y. Yoshikawa, K. Aoki, N. Ichinose, Large-size β-Ga2O3 single crystals and wafers, J. Cryst. Growth 270 (2004) 420–426.
       http://dx.doi.org/10.1016/j.jcrysgro.2004.06.027

[4]   T. Onuma, S. Fujioka, T. Yamaguchi, Y. Itoh, M. Higashiwaki, K. Sasaki, T. Masui, T. Honda, Polarized Raman spectra in β-Ga2O3 single crystals, J. Cryst. Growth 401 (2014) 330–333.
       http://dx.doi.org/10.1016/j.jcrysgro.2013.12.061

[5]   L. Kong, J. Ma, C. Luan, W. Mi, Y. Lv, Structural and optical properties of heteroepitaxial beta Ga2O3 films grown on MgO (100) substrates, Thin Solid Films, 520 (2012) 4270–4274.
       http://dx.doi.org/10.1016/j.tsf.2012.02.027

[6]   K. Sasaki, M. Higashiwaki, A. Kuramata, T. Masui, S. Yamakoshi, J. Cryst. Growth 392 (2014) 30–33.
       http://dx.doi.org/10.1016/j.jcrysgro.2014.02.002

[7]   P.J. Pallister, S.C. Buttera, S.T. Barry, Self-seeding gallium oxide nanowire growth by pulsed chemical vapor deposition, Phys. Status Solidi A 212 (2015) 1514–1518.
http://dx.doi.org/10.1002/pssa.201532275

[8]   C.J. Szwejkowski, N.C. Creange, K. Sun, A. Giri, B.F. Donovan, C. Constantin, P.E. Hopkins, Size effects in the thermal conductivity of gallium oxide (β-Ga2O3) films grown via open-atmosphere annealing of gallium nitride, J. Appl. Phys. 117 (2015) 084308.
http://dx.doi.org/10.1063/1.4913601

[9]   F. Lawrenz, P. Lange, N. Severin, J.P. Rabe, C.A. Helm, S. Block, Morphology, Mechanical Stability, and Protective Properties of Ultrathin Gallium Oxide Coatings, Langmuir 31 (2015) 5836–5842.
http://dx.doi.org/10.1021/acs.langmuir.5b00871

[10]   N.V. Golubev, E.S. Ignateva, V.N. Sigaev, A. Lauria, L. De Trizio, A. Azarbod, A. Paleari, R. Lorenzie, Diffusion-driven and size-dependent phase changes of gallium oxide nanocrystals in a glassy host, Phys. Chem. Chem. Phys. 17 (2015) 5141-5150.
http://dx.doi.org/10.1039/C4CP05485G

[11]   W. Xu, H. Cao, L. Liang, J.B. Xu, Aqueous Solution-Deposited Gallium Oxide Dielectric for Low-Temperature, Low-Operating-Voltage Indium Oxide Thin-Film Transistors: A Facile Route to Green Oxide Electronics, ACS Appl. Mater. Interfaces. 7 (2015) 14720-14725.
http://dx.doi.org/10.1021/acsami.5b02451

[12]   H. Cai, H. Liu, H. Zhu, P. Shao, C. Hou, Capacitive Behavior of Single Gallium Oxide Nanobelt, Materials 8 (2015) 5313-5320.
http://dx.doi.org/10.3390/ma8085244

[13]   R. Yuvakkumar, J. Suresh, A. Joseph Nathanael, M. Sundrarajan, S.I. Hong, Novel green synthetic strategy to prepare ZnO nanocrystals using rambutan (Nephelium lappaceum L.) peel extract and its antibacterial applications, Mater. Sci. Eng. C 41 (2014) 17-27.
http://dx.doi.org/10.1016/j.msec.2014.04.025

[14]   R. Yuvakkumar, J. Suresh, A. Joseph Nathanael, M. Sundrarajan, S.I. Hong, Rambutan (Nephelium lappaceum L.) peel extract assisted biomimetic synthesis of nickel oxide nanocrystals, Mat. Lett. 128 (2014) 170-174.
http://dx.doi.org/10.1016/j.matlet.2014.04.112

[15] R. Yuvakkumar, J. Suresh, B. Saravanakumar, A. Joseph Nathanael, S.I. Hong, V. Rajendran, Rambutan peels promoted biomimetic synthesis of bioinspired zinc oxide nanochains for biomedical applications, Spectrochim. Acta Part A. 137 (2015) 250-258.
http://dx.doi.org/10.1016/j.saa.2014.08.022

[16] R. Yuvakkumar, A. Joseph Nathanael, S.I. Hong, Inorganic complex intermediate Co3O4 nanostructures using green ligation from natural waste resources, RSC Adv. 4 (2014) 44495–44499.
http://dx.doi.org/10.1039/C4RA07646J

[17] R. Yuvakkumar, S.I. Hong, Incubation and aging effect on cassiterite type tetragonal rutile SnO2 nanocrystals, J. Mater. Sci. - Mater. Electron. 26 (2015) 2305-2310.
http://dx.doi.org/10.1007/s10854-015-2684-1

[18] R. Yuvakkumar, J. Suresh, B. Saravanakumar, A. Joseph Nathanael, V. Rajendran, S.I. Hong, An environment benign biomimetic synthesis of mesoporous NiO concentric stacked doughnuts architecture, Microporous Mesoporous Mater. 207 (2015) 185–194.
http://dx.doi.org/10.1016/j.micromeso.2015.01.027

[19] R. Yuvakkumar, A. Joseph Nathanael, S.I. Hong, Nd2O3: Novel synthesis and characterization, J. Sol-Gel Sci. Technol. 73 (2015) 511-517.
http://dx.doi.org/10.1007/s10971-015-3629-0

[20] R. Yuvakkumar, S.I. Hong, Structural, compositional and textural properties of monoclinic α-Bi2O3 nanocrystals, Spectrochim. Acta Part A. 144 (2015) 281-286.
http://dx.doi.org/10.1016/j.saa.2015.02.093

[21] R. Yuvakkumar, S.I. Hong, Structural phase transitions in niobium oxide nanocrystals, Phase Transitions 88 (2015) 897-906.
http://dx.doi.org/10.1080/01411594.2015.1033420

[22] L. Binet, D. Gourier, Origin of the blue luminescence of β-Ga2O3, J. Phys. Chem. Solids 59 (1998) 1241-1249.
http://dx.doi.org/10.1016/S0022-3697(98)00047-X

[23] T. Harwig, F. Kellendonk, Some observations on the photoluminescence of doped β-galliumsesquioxide, J. Solid State Chem. 24 (1978) 255-263.
http://dx.doi.org/10.1016/0022-4596(78)90017-8

CHAPTER 3

Structural, optical and magnetic properties of $Ga_{2-x}Fe_xO_3$

M. Charles Robert[1], S. Sasikumar[2]*, S. Saravanakumar[3], R. Saravanan[2]

[1]Department of Physics, HKRH College, Uthamapalayam-625 533, Tamil Nadu, India

[2]Research Centre and Post Graduate Department of Physics, The Madura College, Madurai – 625 011, Tamil Nadu, India

[3]Department of Physics, Kalasalingam University, Krishnankoil – 626 126, Tamil Nadu, India
Email: jothycharles@gmail.com; sasikuhan@gmail.com*; saravanaphysics@gmail.com

Abstract

The present work focuses on the multiferroic gallium iron oxide compound which could be a good candidate for magnetic storage applications. In this study, the poly crystalline samples of $Ga_{2-x}Fe_xO_3$ (x=0.8, 1.0, 1.2) were synthesized using the standard solid state reaction method. The prepared samples were characterized by powder X- ray diffraction and the powder profile pattern was studied using the Rietveld refinement technique. The average crystallite sizes were evaluated using the Debye Scherrer's formula and found to be in the range of nanometers. The magnetic properties were studied by using the vibrating sample magnetometer (VSM). The surface morphologies were analyzed by scanning electron microscopy (SEM). The elements present in the sample were verified by using EDS spectra. The band gap energy of $Ga_{2-x}Fe_xO_3$ was evaluated using UV-Visible data.

Keywords

Rietveld Refinement, Vibrating Sample Magnetometer, Scanning Electron Microscopy

Contents

1.   Introduction

$GaFeO_3$ crystallizes in non-centrosymmetric orthorhombic structure ($Pna2_1$) while α-Fe_2O_3 crystallizes in the corundum structure ($R3c$), and Ga_2O_3 has the stable monoclinic structure. The term multi ferroics has been used to describe materials in which two or all three phenomena: ferroelectricity, ferromagnetism (FM), and ferroelasticity occur in the same phase or in the multiphase composite [1]. Multiferroics emerge as materials for real life applications in microelectronics, spintronics, data storage, and the computing hardware industry [2-5]. This class of materials can find numerous industrial applications [6] such as a new generation of RAMs (Random Access Memories). The concept of magneto electric random access memory (MERAM) which requires a coupling between electric and magnetic properties for data storage has appeared [7, 8]. Such a device combines the advantages of the magnetic random access memories (MRAMs) in terms of access time and endurance with those of the ferroelectric random access memories (FeRAMs) in terms of writing energy. The MERAMs represent consequently a considerable interest concerning the data storage. These technological advances in terms of energy saving, endurance and miniaturization are in excellent agreement with the environmental concerns. While most of the [9-14] research on multiferroic and magnetoelectric materials has focused on $BiFeO_3$ and its limitations necessitate the need to examine other materials such as Gallium ferrite ($GaFeO_3$). There are indications in the earlier literature that both $GaFeO_3$ and $AlFeO_3$ are piezoelectric [15, 16], and a recent study reports magneto dielectric effect in these oxides [17]. What is not clear, however, is whether $GaFeO_3$ and $AlFeO_3$ are ferroelectric, and if so whether they are multiferroic. They are in fact, ferrimagnetic [18, 19], with interesting dielectric properties. The Curie temperature T_C of $GaFeO_3$ is 210K. GFO is a room temperature piezoelectric and a ferrimagnet with transition temperature close to room temperature and has been shown to possess significant linear magneto electric coupling at low temperatures [19]. The present study of the $Ga_{2-x}Fe_xO_3$ system was undertaken to investigate the effect of the substitution of Fe on the structure, magnetic and optical properties of Ga_2O_3.

2.   Sample preparation

Bulk samples of $Ga_{2-x}Fe_xO_3$ were synthesized via the standard solid state reaction route, where first of all appropriate amounts of Ga_2O_3 (Alfa Aesar 99.99%) and Fe_2O_3 (Alfa Aesar 99.99%) were weighted to their stoichiometric molar ratio and ground in an agate mortar. After this, the resulting material was pelletized. The resulting pellets were sintered at 1350°C for 10 h in air followed by furnace cooling.

3.   Results and discussion

3.1   Powder X-ray analysis

The synthesized $Ga_{2-x}Fe_xO_3$ (x=0.8, 1.0, 1.2) bulk samples were characterized using powder X-ray (CuKα) diffraction (Philips, X-PERT PRO, and Netherland) at the National Institute of Interdisciplinary Science and Technology (NIIST), Trivandrum, Kerala, India. An angular range of 2θ from 10^0 to 120^0 with a step width of 0.02° was scanned. The probing X-ray beam was CuKα with λ = 1.5406Å. The XRD patterns of the sintered powders of all compositions are shown in fig. 1(a) which is compatible with the JCPDS data of orthorhombic GaFeO$_3$ structure (JCPDS 26-0673). The positions of XRD peaks match perfectly with the JCPDS and no additional peaks due to other phases are observed. The EDS microanalysis did not detect any solid solubility, which confirms the results obtained with X-ray analysis. The characteristics of the enlarged XRD patterns are shown in fig. 1b. Moreover, it can be seen that the diffraction peaks of the powder pattern shift to higher angles with increasing x, which indicates that the lattice parameters have a small shrinkage owing to the relatively smaller ionic radius of Fe^{3+} (0.55Å) compared to that of Ga^{2+} (0.62Å) [20]. The lattice parameters obtained from XRD are shown in table 1.

3.2   Rietveld refinement

GaFeO$_3$ belongs to an orthorhombic structure with a non-centrosymmetric space group ($Pna2_1$). Rietveld refinement [21] carried out by JANA2006 [22] software. The JANA2006 software can effectively characterize the material which contain anisotropy of size and strain which causes the diffraction profiles to broaden. Rietveld refinement [21] gives the automatic refinements profile for the crystalline materials. The structural parameters obtained from refinement are tabulated in Table 1. The Rietveld refinement profiles for $Ga_{2-x}Fe_xO_3$ system are shown in fig. 2(a-c).

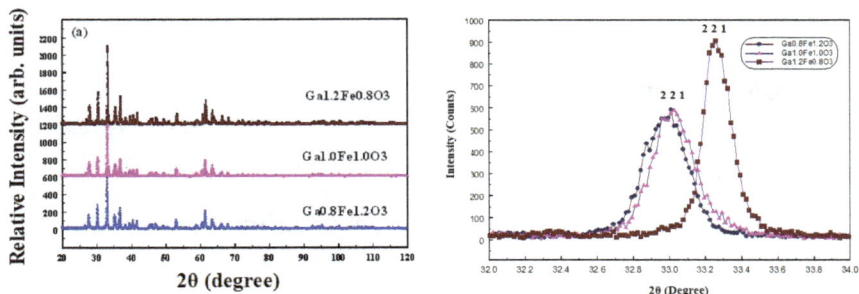

Figure. 1(a). X-diffraction patterns for $Ga_{2-x}Fe_xO_3$ (b) enlarged (221) peak.

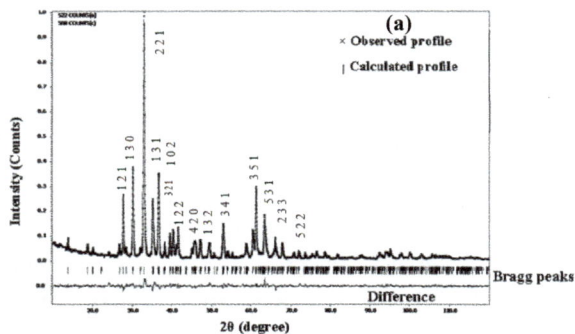

Figure. 2 (a) Rietveld refinement profile for $Ga_{0.8}Fe_{1.2}O_3$.

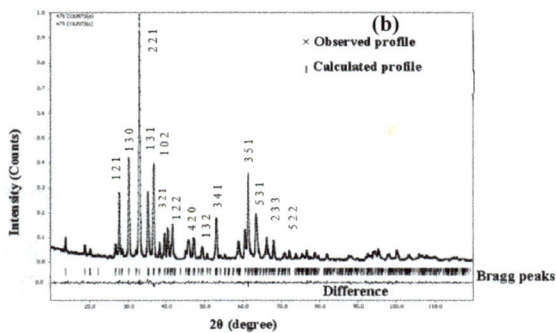

Figure. 2 (b) Rietveld refinement profile for $Ga_{1.0}Fe_{1.0}O_3$.

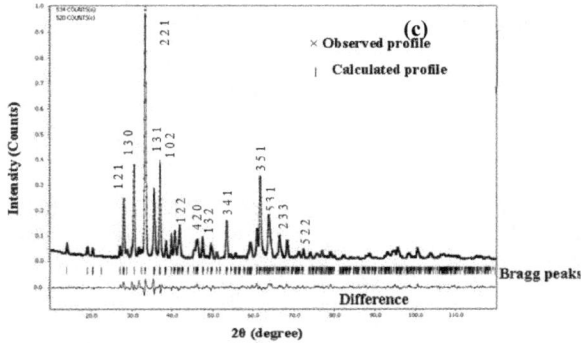

Figure. 2 (c). Rietveld refinement profile for $Ga_{1.2}Fe_{0.8}O_3$.

Table 1. Structural parameters for $Ga_{2-x}Fe_xO_3$ from JANA2006 software.

Parameters	$Ga_{0.8}Fe_{1.2}O_3$	$Ga_{1.0}Fe_{1.0}O_3$	$Ga_{1.2}Fe_{0.8}O_3$
a(Å)	8.6949 (2)	8.7364 (5)	8.7422 (9)
b(Å)	9.3380(6)	9.3866(2)	9.3907(7)
c(Å)	5.0540(5)	5.0740(4)	5.0851(1)
$\alpha = \beta = \gamma$	90°	90°	90°
Density (gm/cc)	5.6208	5.4508	5.0851
Volume(Å^3)	410	416	417
R_{obs} (%)	1.10	3.16	2.46
R_p (%)	4.65	6.95	7.26
$F_{(000)}$	648	640	656
GOF	0.29	0.41	0.45

3.3   Scanning electron microscopy and EDS

The surface morphology of the samples was analyzed by scanning electron microscopy (SEM) equipped with energy dispersive spectrometer (EDS). The particle sizes of the $Ga_{2-x}Fe_xO_3$ polycrystalline systems were characterized using SEM (scanning electron microscopy) at Karunya University, Coimbatore, India. The analysis on the range of the

25

particle sizes of the prepared samples was done using the scanning electron microscope model JEOL JSM 6390. Fig. 3(a-c) shows the SEM micrographs of the samples. From SEM micrographs, it is clearly observed that the particle sizes are found to be in the range of 0.5μm. For determining the compositions of transition metals in $Ga_{2-x}Fe_xO_3$ bulk samples, X-ray dispersive spectroscopy (EDS) analysis has been used. EDS spectra of $Ga_{2-x}Fe_xO_3$ (x=0.8, 1.0, and 1.2) samples are shown in fig. 3(d-f).

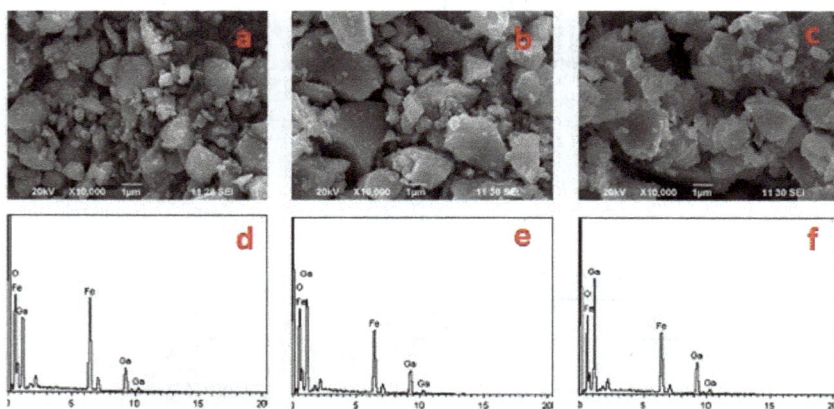

Figure. 3   SEM micrographs of the prepared $Ga_{2-x}Fe_xO_3$, a) x=0.8 b) x=1.0 c) x=1.2 ceramics with their elemental composition found using EDS, (d) x=0.8 (e) x=1.0 (f) x=1.2.

3.4   Magnetic property of $Ga_{2-x}Fe_xO_3$

To study the origin of the magnetic behavior in the $Ga_{2-x}Fe_xO_3$ powders, we performed an experimental analysis using vibrating sample magnetometer at room temperature. The synthesized gallium iron oxide samples were characterized by vibrating sample magnetometer at Sophisticated Analytical Instrument facility (SAIF), IIT Chennai, India. A graph is plotted between the magnetic field (G) and the magnetic moment. Fig. 4 shows the magnetization curve of $Ga_{2-x}Fe_xO_3$ and it gives a clue on the magnetic property of the gallium iron oxide powders. The hysteresis curve for $Ga_{0.8}Fe_{1.2}O_3$ composition is shown in fig.4. From fig.4 we can conclude that $Ga_{0.8}Fe_{1.2}O_3$ exhibits a ferromagnetic property and the hysteresis is not so prominent for the other two concentrations which may be paramagnetic.

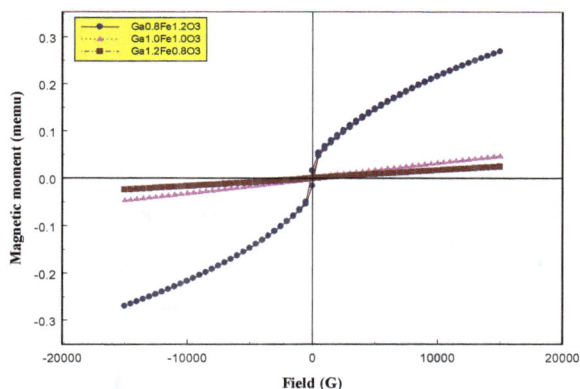

Figure. 4 Magnetization curve for $Ga_{2-x}Fe_xO_3$.

Table 2. Parameter of $Ga_{2-x}Fe_xO_3$ form VSM measurements.

Samples	H_c (G)	M_s(emu)	M(g) (10^{-3})	M_r(emu) (10^{-6})
$Ga_{0.8}Fe_{1.2}O_3$	122.96	0.2692	64.00	16206.0
$Ga_{1.0}Fe_{1.0}O_3$	96.24	0.0469	84.00	582.97
$Ga_{1.2}Fe_{0.8}O_3$	146.66	0.0244	74.00	379.94

3.5 UV-Visible spectra for $Ga_{2-x}Fe_xO_3$

In order to study the effect of strain on the energy band gap and absorption peaks, optical absorption spectrographic studies were carried out, by recording the UV-Visible absorption spectrum of the multiferroic compound $GaFeO_3$. The prepared samples were analyzed by UV-Visible spectrum at (Sophisticated Analytical Instrument facility) SAIF, Cochin, India. UV-Visible absorption spectra in the form of $(\alpha h v^2)$ versus hv for all prepared samples (x=0.8, 1.0 and 1.2) are shown in fig. 5. An estimate of the optical band gap is obtained using the following equation for a $GaFeO_3$. According to the equation $(\alpha h v)^2 = A (hv-E_g)$, [23] the band gaps for $Ga_{2-x}Fe_xO_3$ are obtained to be 3.7730 eV, 3.8326 eV, and 3.9038 eV respectively. The band gaps of the materials indicate an insulating property of the material.

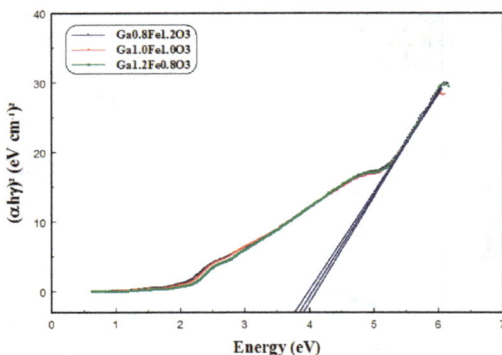

Figure 5. Optical absorption spectra of $Ga_{2-x}Fe_xO_3$ in the form of $(\alpha h v^2)$ versus the photon energy E.

Conclusions

The multiferroic polycrystalline compound $Ga_{2-x}Fe_xO_3$ (x=0.8, 1.0, 1.2) has been synthesized by solid state reaction method. The synthesized samples have been characterized by PXRD. The surface morphology study has been carried out using scanning electron microscopy (SEM) and the particle size of the prepared samples are found to be around 500nm. The elemental compositions have been analyzed by energy dispersive X-ray spectroscopy and it confirms the expected stoichiometry of the prepared samples. Magnetic properties of the gallium iron oxide have been studied using a vibrating sample magnetometer (VSM). A ferromagnetic curve is observed in the sample with higher Fe concentration at the host lattice. The band gap values have been evaluated using data from the UV-Visible spectra.

References

[1]   K. Kelm, W. Mader, Z. Anorg. Allg. Chem, Synthesis and structural analysis of epsilon-Fe_2O_3, 631 (2005) 2383-2389.

[2]   M. E. Lines and A. M. Glass, Principles and Application of Ferroelectrics and Related Materials (Clarendon, Oxford, 1977).

[3]   M. Bibes and A. Barthelemy, Multiferroics: Towards a magnetoelectric memory, Nature Mater. 7 (2008) 425-426. http://dx.doi.org/10.1038/nmat2189

[4]   R. Ramesh and N. A. Spaldin, Multiferroics: progress and prospects in thin films Nature mater. 6 (2007) 21-29. http://dx.doi.org/10.1038/nmat1805

[5]   E. Bousquet, M. Dawber, N. Stucki, C. Lichtensteiger, P. Hermet, S. Gariglio, J. M. Triscone, and P. Ghosez, Nature (London) 452, 723 (2008). http://dx.doi.org/10.1038/nature06817

[6]   M. Gajek, M. Bibes, S. Fusil, K. Bouzehouane, J. Fontcuberta, A. Barthelemy, A. Fert, Tunnel junctions with multiferroic barriers. Nature Mater, 6(4) 2007 296–302. http://dx.doi.org/10.1038/nmat1860

[7]   M. Bibes, Multiferroics BA. Towards a magnetoelectric memory. Nature Mater 7(6) 2008 425–426. http://dx.doi.org/10.1038/nmat2189

[8]   V.E. Wood, A.E. Austin, Possible applications for magnetoelectric materials. Int J Magn 5(4) 19743 03–15.

[9]   H. Schmid, Multi-ferroic magnetoelectrics, Ferroelectrics 162 (1994) 317–338. http://dx.doi.org/10.1080/00150199408245120

[10]   J.F. Scott, Data storage: Multiferroic memories, Nat. Mater. 6 (2007) 256–257. http://dx.doi.org/10.1038/nmat1868

[11]   G.A. Prinz, Magnetoelectronics, Science 282 (1998) 1660–1663. http://dx.doi.org/10.1126/science.282.5394.1660

[12]   A. Roy, R. Gupta, A. Garg, Magnetoelectric Memories: A Review, Adv. Cond. Mat. Phys. Article ID 926290 (2012),

[13]   H. Schmid, On the possibility of ferromagnetic, antiferromagnetic, ferroelectric and ferroelastic domain reorientations in magnetic and electric fields, Ferroelectrics 221 (1999) 9–17. http://dx.doi.org/10.1080/00150199908016431

[14]   W. Eerenstein, N.D. Mathur, J.F. Scott, Multiferroic and magnetoelectric materials, Nature 442 (2006) 759–765. http://dx.doi.org/10.1038/nature05023

[15]   J.P. Remeika, $GaFeO_3$: A Ferromagnetic-Piezoelectric Compound, J. Appl. Phys. 31 (1960) S263. http://dx.doi.org/10.1063/1.1984690

[16]   G.T. Rado, Observation and Possible Mechanisms of Magnetoelectric Effects in a Ferromagnet, Phys. Rev. Lett. 13 (1964) 335. http://dx.doi.org/10.1103/PhysRevLett.13.335

[17]   A. Shireen, R. Saha, P. Mandal, A. Sundaresan, C.N.R. Rao, Multiferroic and magnetodielectric properties of the $Al_{1-x}Ga_xFeO_3$ family of oxide, J. Mater. Chem. 21 (2011) 57-59. http://dx.doi.org/10.1039/C0JM02688C

[18]   L.F. Cotica, I.A. Santos, M. Venet, D. Garcia, J.A. Eiras, A.A. Coelho, Dielectric and magnetic coupling in lead-free $FeAlO_3$ magnetoelectric compound Solid State, Commun. 147 (2008) 123-125. http://dx.doi.org/10.1016/j.ssc.2008.05.001

[19]   R.B. Frankel, N.A. Blum, S. Foner, A.J.Freeman, M. Schieber, Ferrimagnetic structure of magnetoelectric $Ga_{2-x}Fe_xO_3$, Phys. Rev. Lett. 15 (1965) 958-960. http://dx.doi.org/10.1103/PhysRevLett.15.958

[20]   R.D. Shannon, Revised effective ionic radii and systematic studies of interatomic distances in halides and chalcogenides. Acta Cryst. A32 (1976) 751-767. http://dx.doi.org/10.1107/S0567739476001551

[21]   H.M. Rietveld, A Profile Refinement Method for Nuclear and Magnetic, J. Appl. Crystallogr. 2 (1969) 65. http://dx.doi.org/10.1107/S0021889869006558

[22]   V. Petricek, M. Dusek and L. Palatinus, (2006) Jana, The crystallographic computing system (Institute of Physics), Praha, Czech Republic.

[23]   J.I. Pancove, (1971). Optical processes in semiconductors. Englewood Cliffs, NJ, USA: Prentice Hall.

CHAPTER 4

Sol-gel synthesis and characterization of samarium and manganese substituted calcium hydroxyapatite, $Ca_{10}(PO_4)_6(OH)_2$

I. Bogdanoviciene, A. Beganskiene, A. Kareiva

Department of Inorganic Chemistry, Vilnius University, Naugarduko 24, LT-03225 Vilnius, Lithuania

Email: aivaras.kareiva@chf.vu.lt

Abstract

An aqueous sol-gel synthesis route has been developed to prepare samarium and manganese substituted calcium hydroxyapatite ($Ca_{10-x}M_x(PO_4)_6(OH)_2$, CHAp:M) samples. The final products were obtained by calcination of the dry Ca(M)-P-O precursor gels for 5 h at 1000 °C. The phase transformations, composition and structural changes in the polycrystalline samples were studied by infrared (IR) spectroscopy, X-ray powder diffraction (XRD) analysis, scanning electron microscopy (SEM), UV-visible reflection spectroscopy and luminescence measurements. It was demonstrated that phase purity, morphology and optical properties of the end products highly depends on the amount of lanthanide or transition metal in CHAp.

Keywords

Calcium Hydroxyapatite, Sol-Gel Processing, Substitution Effect, Samarium, Manganese

Contents

1. Introduction

With an increase of the mean population age, the development and optimization of bone regeneration techniques represent a major clinical need. Ceramics used for the repair and reconstruction of diseased or damaged parts of the musculo-skeletal system, termed bioceramics, may be bioinert (alumina, zirconia), resorbable (tricalcium phosphate), bioactive (hydroxyapatite, bioactive glasses and glass-ceramics), or porous for tissue ingrowth (hydroxyapatite-coated metals, alumina) [1-3]. The specific chemical, structural and morphological properties of CHAp bioceramics are highly sensitive to the changes in chemical composition and processing conditions [4-8]. Most natural apatite is non-stoichiometric because of the presence of minor constituents such as cations (Mg^{2+}, Mn^{2+}, Zn^{2+}, Na^+, Sr^{2+}) or anions (HPO_4^{2-} or CO_3^{2-}) [9-11]. The traces of metal ions introduced in apatite structure can affect the lattice parameters, the crystallinity, dissolution kinetics and other physical properties of apatite. The reports regarding the substitution of Ca^{2+} ions by bivalent or trivalent metal ions attracted attention during the past few years [8, 12–14]. However, according to these reports some metal ions did not enter the crystal lattice of CHAp.

It was also suggested that transition metals and lanthanide elements might play an important role in enamel demineralization reduction [15, 16]. Apatites with the chemical formula $Sr_{7-x}Ca_xLa_3(PO_4)_3(SiO_4)_3F_2$, where x = 0, 1 and 2, were prepared by mechanochemical synthesis using a planetary mill [17]. Cerium (IV)-substituted hydroxyapatite nanoparticles were synthesized by the co-precipitation method from aqueous solutions of various Ce/(Ce+Ca) atomic ratios [18]. The results confirmed that cerium ions in the +4 as well as in the +3 states have been incorporated into the hydroxyapatite lattice. The antibacterial properties of cerium-substituted hydroxyapatite powders against Escherichia coli and Staphylococcus aureus bacteria were found to increase with increasing cerium content, being more effective against E. coli. In [19] nanosized biphasic CHAp/beta-TCP compounds co-doped with Mn^{2+} and Eu^{3+}/Eu^{2+}, and Dy^{3+}/Pr^{3+} ions with sphere and rod-shaped were obtained. These nanoparticles were synthesized via the hydrothermal route and tailored to present red-near infrared persistent luminescence after UV excitation. These nanoparticles have been tested for the first time for in vivo imaging on small animal as proof of concept of prospective highly biocompatible nanoprobes. Nanocrystalline hydroxyapatite has a high loading capacity, therefore it can be used in pharmaceutical science as a carrier for drugs or a system for controlled poorly soluble drug release [20].

Manganese is important element from the point of view of the synthesis of mucopolysacharides and its deficit causes lowering of synthesis of organic matrix and retards endochondral osteogenesis and formation of bone abnormalities as decreasing of thickness or length of bones and their deformation. Manganese influences regulation of bone remodelling because of its low content in the body it is connected with the rise of the concentration of calcium, phosphates and phosphatase in the cells. Manganese insufficiencies in the human body are probably a contributing cause for osteoporosis [21-23]. Recently, the search for phosphors emitting in the visible range has stimulated a growing interest in Sm^{3+}-doped crystalline materials [24, 25]. Particularly, research on Sm^{3+} ion is significant due to its increasing demand in various fluorescent devices, colour displays and temperature sensors [26].

Over the last few decades, the sol-gel technique has been used to prepare a variety of mixed-metal nanoporous oxides, nanomaterials, nanoscale architectures and organic-inorganic hybrids [27-32]. Previous work has shown that sol-gel process offers considerable advantages such as better mixing of the starting materials and excellent chemical homogeneity in the final product. Moreover, the molecular level mixing and the tendency of partially hydrolyzed species to form extended networks facilitate the structure evolution thereby lowering the crystallization temperature. Recently, for the preparation of calcium hydroxyapatite samples with different properties an aqueous sol-gel processing route was elaborated [33-35]. The main aim of this study was to investigate samarium and manganese substitution effects in $Ca_{10-x}(Sm, Mn)_x(PO4)_6(OH)_2$ synthesized using an environmentally friendly aqueous sol-gel method.

2. Experiments

Calcium hydroxyapatite powders substituted by samarium or manganese $Ca_{10-x}(Sm, Mn)_x(PO_4)_6(OH)_2$ having different concentrations of metals (x = 0.01; 0.025; 0.05; 0.1; 0.5 and 1.0) were prepared by aqueous sol-gel method. In the sol-gel process $Ca(CH_3COO)_2 \cdot H_2O$ ($\geq$ 99 %, Fluka), samarium (III) oxide Sm_2O_3 (99.9 %, Merck), manganese (II) acetate tetrahydrate $Mn(CH_3COO)_2 \cdot 4H_2O$ (99 %, AppliChem), nitric acid HNO_3 (63 %, Eurochemical) and phosphoric acid H_3PO_4 (85%, Eurochemical) were selected as starting materials. The tartaric acid $C_4H_6O_6$ (99.5 %, Aldrich) was used as complexing agent. Two different series of the gels were prepared: (1) samarium (III) oxide was dissolved in distilled water containing a small amount of nitric acid at a temperature of 80 $^\circ$C. This solution is cooled to room temperature and mixed with calcium acetate monohydrate aqueous solution; (2) manganese (II) acetate tetrahydrate and calcium acetate monohydrate aqueous solutions were mixed. Next, phosphoric and tartaric acids were added to the above solutions. The resulting solution was vigorously stirred for 12 h at 65 °C. Obtained sol was concentrated under evaporation till the transparent gel has formed. Further, the gel was dried for 24 h at 100 °C and then calcined for 5 h at 800 °C in air. The powders were grinded in agate mortar to increase the homogeneity and additionally heated for 5 h at 1000 °C in air.

X-ray diffraction (XRD) analysis was performed on a Bruker AXE D8 Focus diffractometer with a LynxEye detector using Cu Kα radiation. Infrared spectra of samples in KBr pellets were recorded with a Bruker Equinox 55/S/NIR FTIR spectrometer (resolution 1 cm^{-1}). The particle size and morphology of the resultant $Ca_{10-x}(Sm, Mn)_x(PO_4)_6(OH)_2$ powders were examined using FE-SEM Zeiss Ultra 55 scanning electron microscope with In-Lens detector. UV-Vis diffuse reflectance spectra were recorded on Perkin-Elmer Lambda 35 UV-Vis spectrophotometer with an integrated 50 mm sphere attachment. Investigation of luminescent properties was performed with PerkinElmer LS-55 fluorescence spectrometer

3. Results and discussion

The XRD patterns of samarium or manganese substituted compounds $Ca_{10-x}(Sm, Mn)_x(PO_4)_6(OH)_2$ (x = 0.01, 0.025, 0.05, 0.1, 0.5 and 1.0) obtained at 1000 °C are shown in figs. 1 and 2. Apparently, the phase composition of synthesis products depends slightly on the amount of substituent in CHAp powders. The XRD patterns of calcium hydroxyapatite powders, when Ca^{2+} substituted by Sm^{3+} show that the polycrystalline single-phase $Ca_{10-x}Sm_x(PO_4)_6(OH)_2$ was obtained with the lowest levels of Sm^{3+}, namely 0.01 and 0.025 mol%. The samples with higher amount of samarium contain $Ca_{10}(PO_4)_6(OH)_2$ as a main phase, however, the diffraction peaks corresponding to β-$Ca_3(PO4)_2$ (β-TKP) phase are also seen. The intensity of these peaks increases with increasing the amount of samarium from 0.05 to 1.0 mol%.

From fig. 2 it is also seen that phase purity of Mn^{2+} substituted calcium hydroxyapatite samples also depends on the manganese amount. However, even at smallest amount of manganese (Mn= 0.01 mol%), the negligible peaks of β-$Ca_3(PO_4)_2$ phase are present in the XRD pattern. Finally, the single-phase polycrystalline tricalcium phosphate was obtained when Ca^{2+} was replaced by the highest amount of manganese (Mn= 1.0 mol%).

Infrared spectroscopy (IR) is highly sensitive to the impurities and substitutions in the structure of apatite [11, 36]. The IR spectra of CHAp samples doped with Sm^{3+}or Mn^{2+} are shown in the figs. 3 and 4, respectively. The absorption bands of a significant intensity at 1048 cm^{-1} and 1092 cm^{-1} could be attributed to the factor group splitting of the v_3 fundamental vibrational mode of the $PO4^{3-}$ tetrahedral. The bands located at ~962 cm^{-1} and ~570–602 cm^{-1} correspond to the symmetric stretching modes v_1 and antisymmetric bending modes v_4 P–O vibration of the phosphate groups; respectively [11, 37].

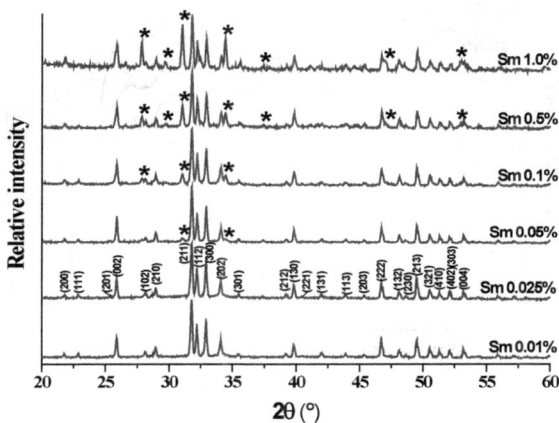

Figure 1.   XRD patterns of Sm^{3+} substituted $Ca_{10-x}Sm_x(PO_4)_6(OH)_2$ with different Sm-concentration obtained at 1000 °C. The peaks of β-$Ca_3(PO_4)_2$ phase are marked (*).

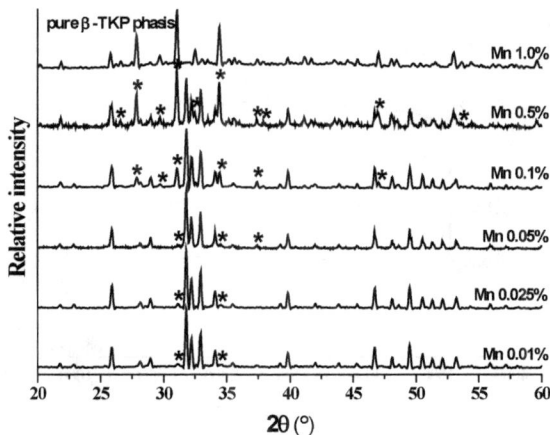

Figure 2.   XRD patterns of Mn^{2+} substituted $Ca_{10-x}Mn_x(PO_4)_6(OH)_2$ with different Mn concentration obtained at 1000 °C. The peaks of β-$Ca_3(PO_4)_2$ phase are marked (*).

35

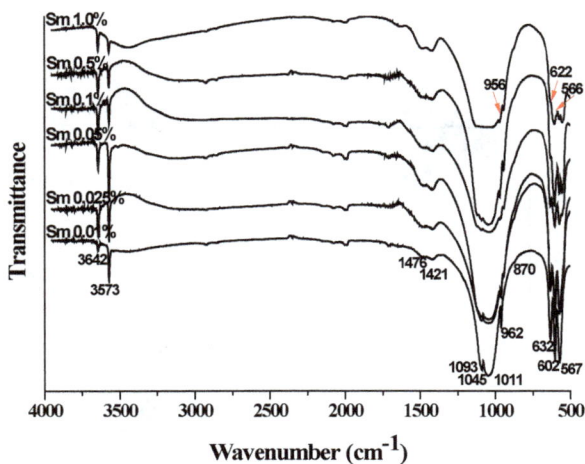

Figure 3.   *IR spectra of Sm^{3+} substituted $Ca_{10-x}Sm_x(PO_4)_6(OH)_2$.*

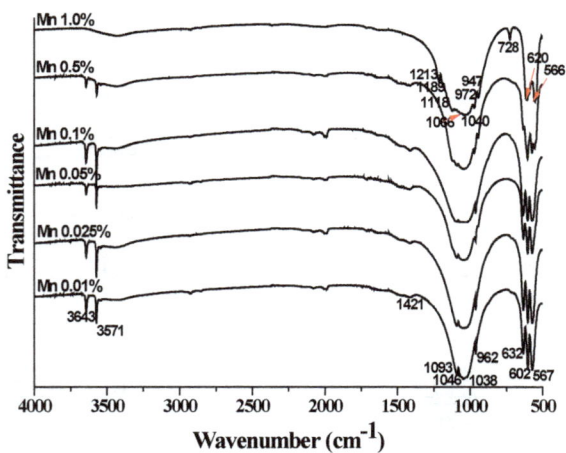

Figure 4.   *IR spectra of Mn^{2+} substituted $Ca_{10-x}Mn_x(PO_4)_6(OH)_2$.*

The peak observed at 630 cm^{-1} in IR spectra is assigned to the O-H group v_L vibrational mode and band at 3573-3644 cm^{-1} is attributed to the stretch vibration mode v_S of OH group in hydroxyapatite structure. The IR spectra also contain the absorption bands characteristic v_3 and v_2 to carbonate located at 1480, 1420and 873 cm^{-1}. Thus, these results let us to conclude that PO_4^{3-} anion is partially substituted by ionic CO_3^{2-} in calcium hydroxyapatite [11, 33].The results of IR spectroscopy are also summarized in Table 1.

Table 1.The results of IR spectroscopy performed on metal substituted Ca$_{10-x}$M$_x$(PO$_4$)$_6$(OH)$_2$ samples.

Band position (cm^{-1})	Assignment
3643	OH$^-$
3571	v_S OH$^-$
632	v_L OH$^-$
1476, 1421	v_3 (CO$_3$)$^{2-}$
870	v_2 (CO$_3$)$^{2-}$
1093,1045, 1011	v_3 (PO$_4$)$^{3-}$
1063	v_3 (PO$_4$)$^{3-}$
962	v_1 (PO$_4$)$^{3-}$
956	v_1 (PO$_4$)$^{3-}$
622, 566	v_4 (PO$_4$)$^{3-}$
602,567	v_4 (PO$_4$)$^{3-}$
505	v_2 (PO$_4$)$^{3-}$
727, 944, 1188, 1213	v_1- v_4 (PO$_4$)$^{3-}$

The representative Ca$_{10-x}$(Sm,Mn)$_x$(PO$_4$)$_6$(OH)$_2$, (x = 0.01, 0.1 and 1.0) samples were investigated by scanning electron microscopy (SEM), from which the grain size and typical morphologies of CHAp could be observed [33, 34, 38]. In our previous studies [33-35], it was demonstrated that the formation of nicely shaped ultrafine elongated particles (microrods and/or microsticks) with about 2-10 μm in length and 1-4 μm in width of CHAp took place when the tartaric acid as complexing agent was used in the sol-gel processing. However, different surface morphology was obtained for the substituted Ca$_{10-x}$(Sm,Mn)$_x$(PO$_4$)$_6$(OH)$_2$ samples prepared using the same synthesis method (fig. 5).

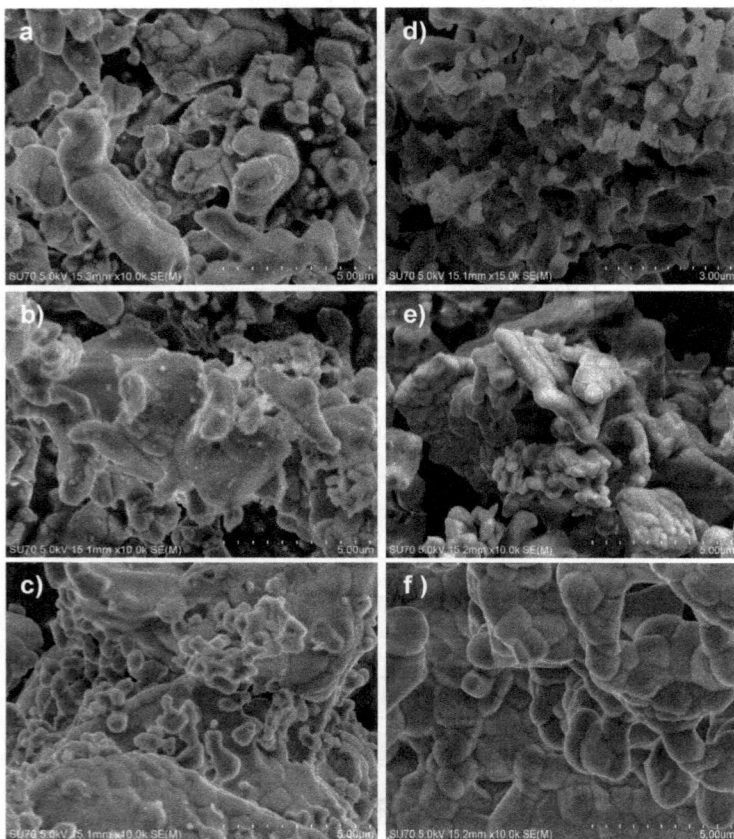

Figure 5. *SEM micrographs of CHAp samples with different Sm^{3+} and Mn^{2+} concentrations: a) Sm^{3+} 0.01%, b) Sm^{3+} 0.1%, c) Sm^{3+} 1.0%, d) Mn^{2+} 0.01%, e) Mn^{2+} 0.1% and f) Mn^{2+} 1,0%.*

As seen from the SEM micrographs, the synthesis products are composed of elongated particles or agglomerates of spherical particles 0.5-5.0 μm in size independent on the nature and amount of substituent metalin CHAp.

The luminescent properties of these samples were also investigated. Fig. 6 shows the reflection, excitation and emission spectra of sol-gel derived samarium substituted CHAp samples.

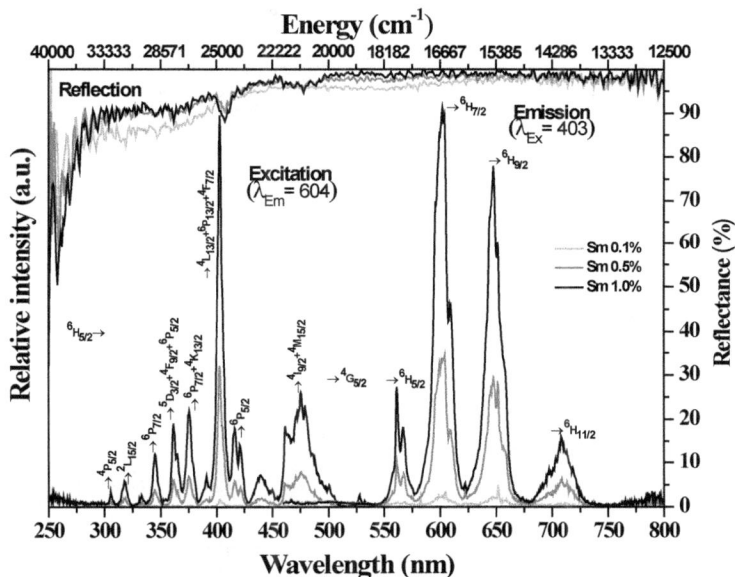

Figure 6.   Reflection, excitation and emission spectra of $Ca_{10-x}Sm_x(PO_4)_6(OH)_2$ samples.

The reflection spectra of $Ca_{10-x}Sm_x(PO_4)_6(OH)_2$ samples contain two main absorption bands located at about 260 nm and 400 nm. The excitation spectra monitored at $\lambda_{em} \approx 604$ nm revealed that the charge transfer (CT) to Sm^{3+} occurs in all hydroxyapatite samples in the UV range (275 nm - 255 nm). In the case of CHAp all observed peaks are due to the excitation from ground state $^6H_{5/2}$ to higher energy levels of Sm^{3+} ion. It is also evident that the CT energy changes monotonically with increasing samarium content in the structure. In addition to the CT band, the excitation spectra of Sm^{3+} doped CHAp samples have also some direct excitation lines. The main is peaked at 403 nm and 455 nm, corresponding to the transitions from ground state to higher energy levels. All powders were excited at 403 nm for taking the emission spectra. The major emission lines are peaked at 604 nm and 652 nm, originating from the excited state level $^4G_{5/2}$ to ground levels $^6H_{5/2}$ and $^6H_{7/2}$, respectively[24, 25].The highest intensity of $^6H_{5/2} \rightarrow {}^4G_{5/2}$ transition was observed for CHAp:1.0% Sm^{3+} specimen.

Fig. 7 shows the reflection and emission spectra of sol-gel derived manganese substituted CHAp samples.

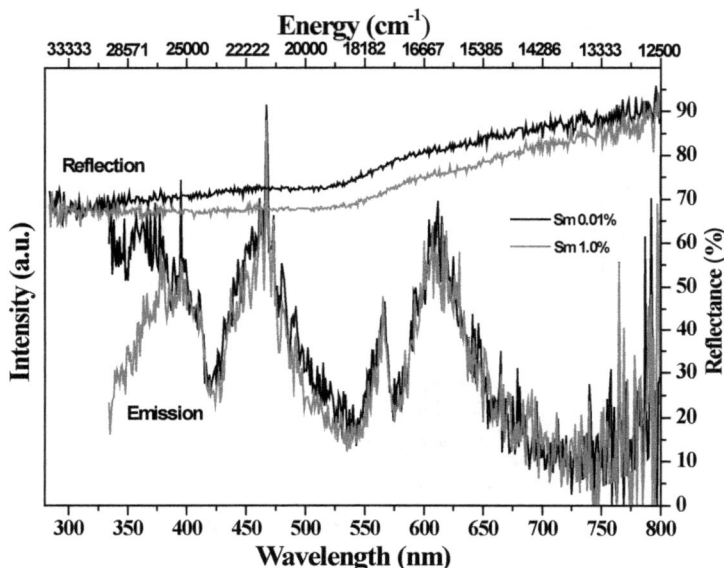

Figure 7.   *Reflection and emission spectra of $Ca_{10-x}Mn_x(PO_4)_6(OH)_2$ samples.*

As seen, the reflection spectra is wavelength-independent in the whole investigated range (250-800 nm). Moreover, the reflection spectra of $Ca_{10-x}Mn_x(PO_4)_6(OH)_2$ samples do not contain any absorption bands in this region. Consequently, the emission spectra do not show any luminescence and contains only noise-like bumps the intensity and position of which are independent on the manganese concentration.

Conclusions

Samarium- and manganese-substituted calcium hydroxyapatite $(Ca_{10-x}Sm_x(PO_4)_6(OH)_2$, $Ca_{10-x}Mn_x(PO_4)_6(OH)_2)$ powders were synthesized using sol–gel method for the first time to our best knowledge. The results of X-ray diffraction analysis showed the formation of almost single CHAp phase at low concentrations of samarium and manganese($x = 0.01$; 0.025; 0.05; 0.1; 0.5 and 1.0). It was demonstrated that infrared spectroscopy is very effective method to characterize the formation of lanthanide- and transition metal-substituted CHAp. The bands due to the factor group splitting of the v_3 fundamental vibrational mode of the PO_4^{3-}, symmetric stretching modes v_1 and antisymmetric bending modes v_4 P–O vibration of the phosphate groups have been determined in FTIR spectra of synthesized materials. The bands assigned in hydroxyapatite spectra to the O-H group vibrational modes were also evident. The SEM results showed that $Ca_{10-x}Sm_x(PO_4)_6(OH)_2$ and $Ca_{10-x}Mn_x(PO_4)_6(OH)_2$ solids were homogeneous having small particle size distribution. As was seen from the SEM micrographs, the synthesis products were composed of elongated particles or agglomerates of spherical particles 0.5-5.0 μm in size independent on the nature and amount of substituent in CHAp. The photoluminescent properties of these samples were also investigated. The PL spectra were definitely attributed and discussed. In particular the excitation spectra of Sm^{3+} doped CHAp were dominated by a broad charge transfer band peaked at 403 nm and 455 nm, while the emission spectra presented the typical $^6H_{5/2} \rightarrow {}^4G_{5/2}$ transitions of Sm^{3+} ions, with maximum intensity for at 629 nm. Among these samples, the highest PL intensity was observed for the CHAp:0.025% Eu^{3+} sample. In the case of Tm^{3+} doped CHAp the excitation spectrum is dominated by the direct excitation $^3H_6 \rightarrow {}^1D_2$ transition peaked at 358 nm, while the emission is characterized by the $^1D_2 \rightarrow {}^3F_4$ transition in the blue region peaked at 604 nm. The highest PL intensity was observed for the CHAp:1.0% Sm^{3+} specimen. In conclusion, these results suggest that Sm substituted CHAp samples showed interesting luminescent properties and could be good candidates for biocompatible drug carriers. However, the manganese substituted $Ca_{10-x}Mn_x(PO_4)_6(OH)_2$ samples did not show any luminescence. On the other hand, manganese-substituted CHAp could be applied for the regulation of bone remodelling and possibly to prevent osteoporosis.

Acknowledgements

This research was funded by a grant KALFOS (No. LJB-2/2015) from the Research Council of Lithuania.

References

[1]   D. Green, D. Walsh, S. Mann, R.O.C. Oreffo. Bone.30 (2002) 810.
       http://dx.doi.org/10.1016/S8756-3282(02)00727-5

[2]   A.C. Lawson, J.T. Czernuszka, Proc. Instr. Mech. Eng. 212 (1998) 413.
       http://dx.doi.org/10.1243/0954411981534187

[3]   A. Anwar, M.N. Asghar, Q. Kanwal, M. Kazmi, A. Sadiqa, J. Molec. Struct. 1117 (2016) 283. http://dx.doi.org/10.1016/j.molstruc.2016.03.061

[4]   A. Bigi, E. Boanini, K. Rubini, J. Solid State Chem. 177 (2004) 3092. http://dx.doi.org/10.1016/j.jssc.2004.05.018

[5]   J. Liu, K. Li, H. Wang, M. Zhu, H. Yan, Chem. Phys. Lett. 396 (2004) 429. http://dx.doi.org/10.1016/j.cplett.2004.08.094

[6]   C.E. Fowler, M. Li, S. Mann, H.C. Margolis, J. Mater. Chem. 15 (2005) 3317. http://dx.doi.org/10.1039/b503312h

[7]   G. Goller, F.N. Oktar, S. Agathopoulos, D.U. Tulyaganov, J.M.F. Ferreira, E.S. Kayali, I. Peker, J. Sol-Gel Sci. Technol. 37 (2006) 111. http://dx.doi.org/10.1007/s10971-006-6428-9

[8]   J. Trinkunaite-Felsen, A. Prichodko, M. Semasko, R. Skaudzius, A. Beganskiene, A Kareiva, Adv. Powder Technol. 26(2015) 1287. http://dx.doi.org/10.1016/j.apt.2015.07.002

[9]   I. Mayer, J.D.B. Featherstone, J. Cryst. Growth. 219 (2000) 98. http://dx.doi.org/10.1016/S0022-0248(00)00608-4

[10]   S. Ben Abdelkader, I. Khattech, C. Rey, M. Jemal, Thermochim. Acta. 376 (2001) 25. http://dx.doi.org/10.1016/S0040-6031(01)00565-2

[11]   E. Garskaite, K. A. Gross, S. W. Yang, Thomas C. K. Yang, J. C. Yang, A. Kareiva, Cryst. Eng. Commun. 16 (2014) 3950. http://dx.doi.org/10.1039/c4ce00119b

[12]   A. Serret, M.V. Cabanas, M. Vallet-Regi, Chem. Mater. 12 (2000) 3836. http://dx.doi.org/10.1021/cm001117p

[13]   I. Bogdanoviciene, M. Cepenko, R. Traksmaa, A. Kareiva, K. Tõnsuaadu. J. Therm. Anal. Calorim. 121 (2015) 107. http://dx.doi.org/10.1007/s10973-015-4507-2

[14]   G. Ciobanu, A.M. Bargan, C. Luca, JOM. 67 (2015) 2534. http://dx.doi.org/10.1007/s11837-015-1467-8

[15]   F.B. Bagam Bisa, H.F . Kappert, W. Schili, J. Oral. Maxillofac. Surg.52 (1994)52. http://dx.doi.org/10.1016/0278-2391(94)90015-9

[16]   I. Bogdanoviciene, M. Misevicius, A. Kareiva, K.A. Gross,Thomas C.K. Yang, G.-T. Pan, H.-W. Fang, J.-C. Yang, Adv. Sci. Technol. 86 (2013) 22. http://dx.doi.org/10.4028/www.scientific.net/AST.86.22

[17] K. Boughzala, K. Bouzouita, Comptes Rendus Chim. 18 (2015) 858.
      http://dx.doi.org/10.1016/j.crci.2014.11.011

[18] G. Ciobanu, A. M. Bargan, C. Luca, Ceram. Int. 41 (2015) 12192.
      http://dx.doi.org/10.1016/j.ceramint.2015.06.040

[19] C. Rosticher, B. Viana, T. Maldiney, C. Richard, C. Chaneac, J. Lumin. 170
      (2016) 460. http://dx.doi.org/10.1016/j.jlumin.2015.07.024

[20] J. Kolmas, S. Krukowski, A. Laskus, M. Jurkitewicz, Ceram. Int. 42 (2016) 2472.
      http://dx.doi.org/10.1016/j.ceramint.2015.10.048

[21] L. Medvecky, R. Stulajterova, L. Parilak, J. Trpcevska, J. Durisin, S.M. Barinov,
      Colloids Surf. A: Physicochem. Eng. Aspects.281 (2006) 221.
      http://dx.doi.org/10.1016/j.colsurfa.2006.02.042

[22] E. Boanini, M. Gazzano, A. Bigi, Acta Biomaterialia.6 (2010) 1882.
      http://dx.doi.org/10.1016/j.actbio.2009.12.041

[23] D. Bani, A. Bencini, M.C. Bergonzi, A.R. Bilia, C. Guccione, M. Severi, R.
      Udisti, B. Valtancoli, J. Pharmaceut. Biomed. Anal. 106 (2015) 197.
      http://dx.doi.org/10.1016/j.jpba.2014.11.021

[24] S. Sakirzanovas, A. Katelnikovas, D. Dutczak, A. Kareiva, T. Jüstel, J. Lumin.132
      (2012) 141. http://dx.doi.org/10.1016/j.jlumin.2011.08.011

[25] A. Stanulis, A. Katelnikovas, D. Enseling, D. Dutczak, S. Sakirzanovas, M. Van
      Bael, A. Hardy, A. Kareiva, T. Jüstel, Opt. Mater. 36 (2014) 1146.
      http://dx.doi.org/10.1016/j.optmat.2014.02.018

[26] S. Chahar, V.B. Taxak, M. Dalal, S. Singh, S.P. Khatkar, Mater. Res. Bull. 77
      (2016) 91. http://dx.doi.org/10.1016/j.materresbull.2016.01.027

[27] J. Livage, M. Henry, C. Sanchez,Progr. Solid State Chem.18 (1988)259.
      http://dx.doi.org/10.1016/0079-6786(88)90005-2

[28] B.L. Cushing, V.L. Kolesnichenko, C.J. O'Connor, Chem. Rev.104 (2004)3893.
      http://dx.doi.org/10.1021/cr030027b

[29] J.D. Mackenzie, E.P. Bescher, Acc. Chem.Res.40 (2007)810.
      http://dx.doi.org/10.1021/ar7000149

[30] C. Yu, D. Cai, K. Yang, J.C. Yu, Y. Zhou, C. Fan, J. Phys. Chem. Solids. 71
      (2010) 1337. http://dx.doi.org/10.1016/j.jpcs.2010.06.001

[31] A. Kareiva, Mater. Sci. (Medziagotyra).17 (2011) 428.

[32]   T.K. Thirumalaisamy, S. Saravanakumar, S. Butkute, A. Kareiva, R. Saravanan, J. Mater. Sci.: Mater. Electron. 27 (2016) 1920. http://dx.doi.org/10.1007/s10854-015-3974-3

[33]   I. Bogdanoviciene, A. Beganskiene, K. Tõnsuaadu, J. Glaser, H-J. Meyer, A. Kareiva, Mater. Res. Bull. 41 (2006) 1754. http://dx.doi.org/10.1016/j.materresbull.2006.02.016

[34]   I. Bogdanoviciene, K. Tõnsuaadu, A. Kareiva, Polish J. Chem. 83 (2009) 47.

[35]   I. Bogdanoviciene, K. Tõnsuaadu, V. Mikli,I. Grigoraviciute-Puroniene, A. Beganskiene, A. Kareiva. Centr. Eur. J. Chem.8 (2010)1323.

[36]   A. Antonakos, E. Liarokapis, T. Leventouri, Biomater. 28 (2007) 3043. http://dx.doi.org/10.1016/j.biomaterials.2007.02.028

[37]   T.-M.G. Chu, J.W. Halloran, S.J. Hollister, S.E. Feinberg, J. Mater. Sci.: Mater. Med.12 (2001)471. http://dx.doi.org/10.1023/A:1011203226053

[38]   A.C. Tas, F. Aldinger, J. Mater Sci:Mater Med. 16 (2005) 167. http://dx.doi.org/10.1007/s10856-005-5919-5

CHAPTER 5

Intrinsic defects in ZnO nanoparticles synthesized by the sol-gel and combustion techniques

V. P. Singh[a,b], Chandana Rath[a,] *

[a]School of Materials Science and Technology,Indian Institute of Technology, Banaras Hindu University, Varanasi, 221005, India

[b]School of Engineering, Indian Institute of Technology Mandi, Himachal Pradesh, India

*Email: chandanarath@yahoo.com

Abstract

We investigate intrinsic defects in ZnO synthesized through sol-gel and combustion techniques using various spectroscopic techniques such as RAMAN, Photoluminescence and Positron annihilation. Pure wurtzite phase of ZnO shows $E_1(LO)$ mode in Raman spectra indicating the presence of interstitial zinc which decreases with increase in calcination temperature irrespective of synthesis techniques and further supported by Positron life time measurement. Williamson-Hall analysis shows tensile and compressive strain for samples made by sol-gel and combustion techniques, respectively. The photoluminescence study demonstrates broad defect band emission (DBE) in compressive strained sample only. The unusual increase in DBE peak with increase in calcination temperature is discussed.

Keywords

Oxides, Chemical Synthesis, Raman Spectroscopy, Positron Annihilation Spectroscopy, Defects

Contents

1. Introduction

ZnO is a typical optically transparent n-type wide band gap semiconductor (~3.2 eV) at room temperature. It has been used in various applications such as UV light emitting diodes, spin functional devices, gas sensors, solar cells, transparent electronics and surface acoustic wave devices etc. [1,2]. Considerable effort has been made to synthesize ZnO with high purity [3, 4]. In fact, these defects create electronic states in the band gap of ZnO which affects its optical and electronic properties [5]. These defects could be zinc interstitials (Zn_i), oxygen interstitials (O_i), zinc vacancies (V_{Zn}) or oxygen vacancies (V_O, V_O^+, V_O^{2+}) [6-8]. In our previous work, we have synthesized ZnO through the coprecipitation route and have shown hydrogen as an impurity attached to V_{Zn} in ZnO lattice [9]. Other methods like sol-gel [10, 11], combustion synthesis [12], hydrothermal synthesis [13], have reported the synthesis of pure ZnO. Moreover, sol–gel technique possesses low processing temperature, ease of composition control, good homogeneity and it is cost effective. However, large scale production of ZnO powder through sol-gel method is time consuming [14]. Alternatively, the solution combustion method generally produces powders "ready to use" with high production rate [12]. The combustion process is an exothermic reaction in between metal nitrates and fuel at low temperatures. The fuels can be citric acid ($C_6H_8O_7$) [15], carbohydrazide (CH_6N_4O), urea (CH_4N_2O) [16], glycine ($C_2H_5NO_2$) etc. [17, 18]. The fuel ignition provides sufficient energy to activate the main reaction rapidly. Usually, a high-quality fuel reacts nonviolently, produce nontoxic gases and acts as a chelating material for the metal cations. During the reaction, metal nitrates act as cation source and as an oxidant. In this process, the reaction is self-sustainable as it holds high-temperatures. The reaction occurs at fast heating rate and therefore needs a short time for the reaction. It has been largely investigated and extensively employed in the large-scale production of the nano sized ZnO powder [12]. Moreover, the combustion method is reliable and cost effective.

The equal importance of sol-gel as well as the combustion techniques demands to understand the defects occurred in ZnO. In this work, our focus is to examine intrinsic defects present in ZnO synthesized through above techniques using various spectroscopic techniques such as XRD, Raman, Photoluminescence and Positron annihilation.

2. Experimental

ZnO powder was synthesized by two different techniques such as sol-gel and solution-combustion. In the sol-gel method, absolute ethanol was used as a solvent, and diethanolamine ($HN(CH_2CH_2OH)_2$, DEA) as a sol stabilizer. 0.5 M molar solutions of Zn was prepared by dissolving zinc acetate 2-hydrate ($Zn(CH_3COO)_2 \cdot 2H_2O$) in 0.1 L of ethanol in a 250 mL conical flask. All materials were of analytical grade. After vigorous stirring of ethanolic solution for 30 min and simultaneous heating at 80 °C, DEA was added drop by drop under constant stirring. The molar ratio of DEA/Zn was fixed at one. In the presence of DEA, the milky solution of zinc acetate became transparent after stirring for 5 min. Further, the solution was continually stirred for 12 hr. The final solution was transparent without any suspension of particles. The transparent organic complex precursor undergoes hydrolysis and inorganic polymerization leading to the formation of sols. By adding water drop wise to the sol with constant stirring, precipitation occurred and a white product was obtained after filtering and washing the precipitate several times with distilled water followed by ethanol. Finally, the product was dried at 250°C for 3 hours.

In the solution-combustion technique, Zn molar solution was prepared by dissolving Zinc nitrate [$Zn(NO_3)_2.6H_2O$] in deionized water. Urea [$CO(NH_2)_2$] dissolved in water was used as a combustion fuel, where its combustion heat was −2.98 kcal/g. All materials were of analytical grade. An appropriate amount of Zinc nitrate and urea solutions were mixed in a beaker and were vigorously stirred under a constant temperature of 80 °C until it became gelatinous. The precursor solution was then poured in a crucible and was introduced in a furnace maintained at 600 °C. Vigorous exothermic combustion reaction between Zinc nitrates and urea took place. Initially, the solution underwent dehydration, spontaneous ignition followed by smoldering combustion with enormous swelling. This process produced foamy and voluminous product followed by large amounts of gases. The whole process took hardly 5 to 6 minutes. The synthesis technique utilized the heat energy released by redox exothermic reaction at a relatively low ignition temperature between metal nitrates and urea. The mechanism of the reaction is given below in eq. (1).

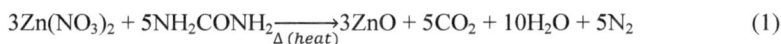

$$3Zn(NO_3)_2 + 5NH_2CONH_2 \xrightarrow{\Delta \ (heat)} 3ZnO + 5CO_2 + 10H_2O + 5N_2 \qquad (1)$$

The dried powder ZnO synthesized through sol-gel further calcined at 600, 800, and 1000°C denoted as S1, S2 and S3, respectively. The ZnO powders synthesized by the solution-combustion route were also calcined at 600, 800, 1000 and 1200°C named as C1, C2, C3 and C4 respectively. All samples were characterized by X-ray diffraction (XRD) using an 18 kW rotating anode (Cu Kα) based Rigaku powder diffractometer. The photoluminescence study was carried out by using a pulsed Nd:YAG laser (Spitlight 600,

Innolas, Germany,) as a light source (355 nm wavelength) with an Ocean Optics QE65000 spectrometer. Raman scattering was performed on a micro-Raman setup of Renishaw, UK equipped with a grating of 1800 lines/mm and a peltier cooled charge coupled device (CCD) detector. The Ar+ laser (514.5 nm) was adopted as the excitation source. A microscope from Olympus (Model: MX50 A/T) was attached with the spectrometer, which focuses the laser light onto the sample. The GRAM-32 software was used for data collection. Positron annihilation lifetime (PAL) measurement was carried out using a fast-fast coincidence system consisting of two 1 in. tapered off BaF_2 scintillators coupled with XP 2020Q photomultiplier tubes. The prompt time resolution of the system using a $_{60}Co$ source with ^{22}Na gate was 298 ps. The lifetime spectra were deconvoluted using the code PATFIT 88 [19].

3.   Results and discussion

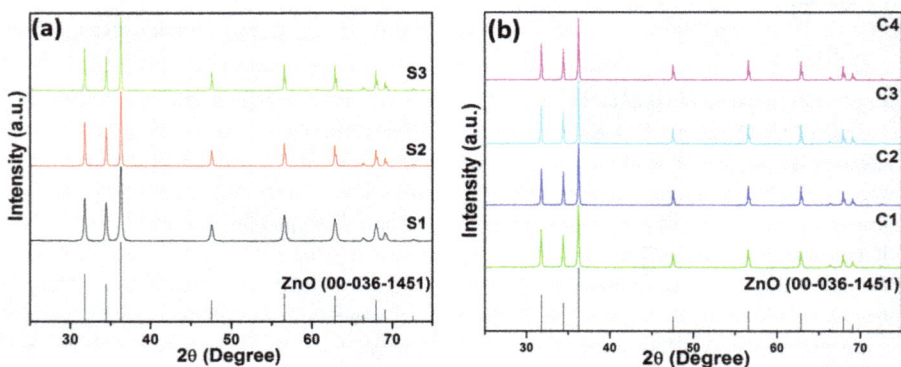

Figure 1.   The XRD patterns of ZnO powder synthesised through (a) Sol-gel route and calcined at different temperatures (S1 = 600, S2 = 800, S3 = 1000 °C) and (b)Combustion route and calcined at different temperatures (C1 = 600, C2 = 800, C3 = 1000, C4 = 1200 °C).

The XRD patterns of ZnO synthesized through sol-gel and solution-combustion techniques calcined at different temperatures, are shown in Fig.1 a & b, respectively. XRD peaks correspond to the wurtzite phase of ZnO (JCPDS 89-1397). No additional peaks are observed indicating no structural changes and/or formation of additional phases in ZnO within the detection limit of XRD.

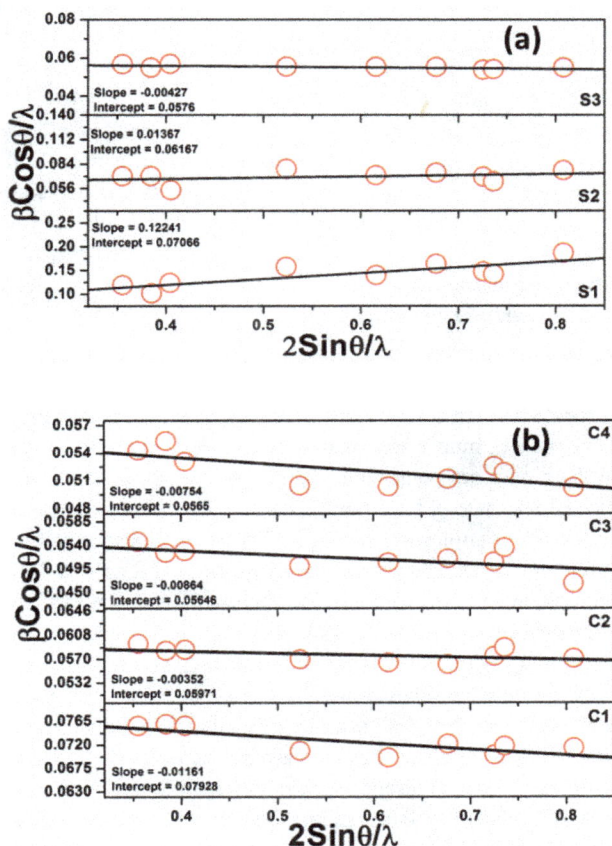

Figure 2. *W-H plot corresponding to XRD patterns of ZnO powder synthesised through (a) Sol-gel route and calcined at different temperatures (S1 = 600, S2 = 800, S3 = 1000 °C), (b)Combustion route and calcined at different temperatures (C1 = 600, C2 = 800, C3 = 1000, C4 = 1200 °C).*

The Williamson-Hall (W-H) plot of the S-series and C-series samples are shown in Fig.2 a & b, respectively. The W-H plot is constructed by choosing $2\sin\theta/\lambda$ as x-axis and $\beta\cos\theta/\lambda$ as Y-axis where β is the full width at half maximum (FWHM) of various Bragg peaks measured after removing the $K\alpha2$ contribution and the instrumental broadening effects, θ is the Bragg angle [20]. For pure particle size broadening, the plot is expected to be a horizontal line parallel to the X-axis whereas, in the presence of strain, it has a non zero slope. The slope as evidenced from the plots in the present case indicates strain in the lattice. The slopes are found to be negative for the samples synthesized by solution combustion route whereas the slopes are positive in sol-gel synthesized samples. Normally in pure and doped ZnO, the slopes obtain from Williamson-Hall (W-H) plots are found to be positive [20]. Negative as well as positive slopes observed by Dutta et al. have been explained as compressive and tensile strain, respectively [21]. Moreover, the strain could be due to the defects originating from some lattice disorder [22-24]. In order to understand the different nature of strain induced in the lattice, we have carried out Raman, Photoluminescence and positron annihilation spectroscopy.

Raman scattering is a versatile technique for fast and non-destructive study of defects and lattice disorders present in the lattice. The Γ point calculation of the brillouin zone for space group C_{6v}^4 of ZnO possess the A_1 and E_1 modes which are both Raman and infrared active, the two nonpolar E_2 modes are Raman active only, and the B_1 modes are inactive. Here the A_1 and E_1 modes are polar which split into longitudinal and transverse optical phonon modes in the presence of long-range Coulomb field. Raman measurements at RT for S-series and C-series samples are shown in Fig. 3a & b, respectively. To assign the Raman modes accurately we have deconvoluted the modes in the range 200-800 cm^{-1} and 800-1300 cm^{-1} and shown in Fig. 4a & b, respectively. The observed Raman modes are well matched with the reported vibrational modes [25-28]. The first order Raman modes such as A_1(TO), E_1(TO), E_2(high), E_1(LO) and multi- phonon modes are present in the experimentally observed modes, shown in Table 1.1. In addition, the relative intensity of E_1(LO) mode at 582 cm^{-1} in S1 and C1 sample is found to be anomalously high and show the gradual decrease in the intensity with increase in calcination temperature. In C-series samples E_1(LO) mode is almost disappeared in C4 (sample annealed at 1200 °C). The origin of the mode is ascribed to the presence of defects such as interstitial zinc or oxygen vacancy in the lattice [29]. In both cases, we have observed decrease in the relative intensity of E_1(LO) mode with increasing calcination temperature as expected. Similar observation has been reported by Sharma et al. [30]. While Zn interstitials (Zn_i) would arise due to the reduced atmosphere during combustion reaction, lack in source of oxidant during the sol-gel synthesis might be responsible for Zn_i. In the case of increasing calcination temperature in air, the excess zinc in the sample is oxidized which reduces E_1(LO) Raman mode [30]. Besides the regular Raman modes, few additional modes at 504, 540, 613 and 657 cm^{-1} have been observed in S4. Similar additional modes at 509/517, 540/536, 617/623 and 699/715 cm^{-1} have been observed in C1 and C4 samples, respectively. In addition to several authors, we have also observed the additional Raman modes in ZnO as well Mg, Co doped ZnO nanoparticles synthesized by co-precipitation technique [31]. Manjon et al. have reported that most of the additional Raman modes observed in doped ZnO samples correspond to the silent modes of ZnO [25]. Serrano et al. have estimated the silent modes theoretically by ab initio calculation [32]. The additional Raman modes observed by us are in agreement with the silent modes reported in literature and compiled in Table 1.1. Although the additional modes in doped ZnO are well reported in literature, their presence in pure ZnO is ambiguous. Strain and $I(E_1(LO))/I(E_2$, High) extracted from W-H plot and Raman analysis decreases with increase in calcination temperature in S-series samples whereas do not show significant variation in C-series samples, shown in Fig. 5. At 1000 °C, both samples bear similar strain.

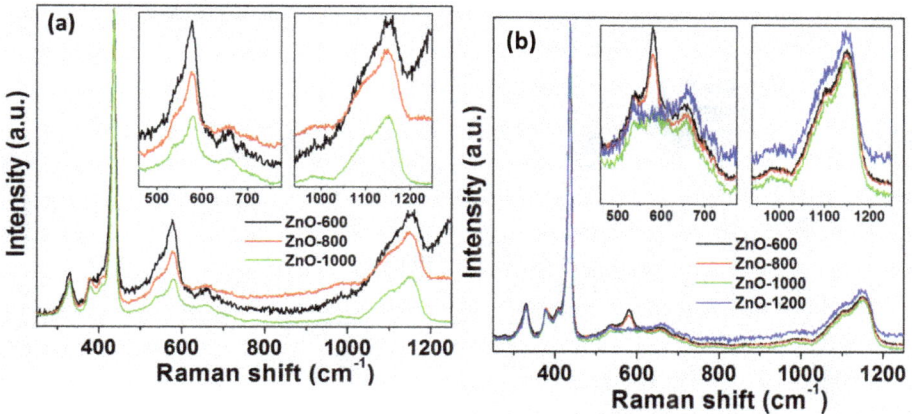

Figure 3.   The Raman spectrum of ZnO powder synthesised through (a) Sol-gel route and calcined at different temperatures (S1 = 600, S2 = 800, S3 = 1000 °C) and (b) Combustion route and calcined at different temperatures (C1 = 600, C2 = 800, C3 = 1000, C4 = 1200 °C).

Table 1. *Reported and observed vibrational modes obtained from Raman spectrum of ZnO synthesized through the sol-gel and the combustion method, calcined at different temperatures.*

Raman vibrational modesofZnO	Reported (cm^{-1})	Sol-gel method (Observed)	Combustion method (Observed)	
		S1 (cm^{-1})	C1 (cm^{-1})	C4 (cm^{-1})
1st order Raman active phonon modes				
$A_1(TO)$	$382^{26, 27}$	381	375	378
$E_1(TO)$	$414^{26, 27}$	413	407	410
E_2^{high}	$439^{26, 27}$	436	437	436
$A_1(LO)$	$574^{26, 27}$	-----	-----	-----
$E_1(LO)$	$580^{26, 27}$	578	582	586
Additional modes (AM)				
$2B_2(low)$	520^{25}	504	509	514
$B_1(high)$	552^{25}	540	540	536
$TA+B_1(high)$	650^{25}	613	617	623
Other	680^{28}	657	699	715
Multi-phonon modes (MP)				
$2E_2$	333^{26}	329	327	329
$TA + LO$	666^{26}	657	658	663
$2TO$	980^{26}	979	984	982
$TO + LO$	1072^{26}	1079	1074	1075
$2LO$	1105^{26}	1107	1102	1104
$2A_1(LO)$, $2E_1(LO)$; 2LO	1158^{26}	1152	1149	1155

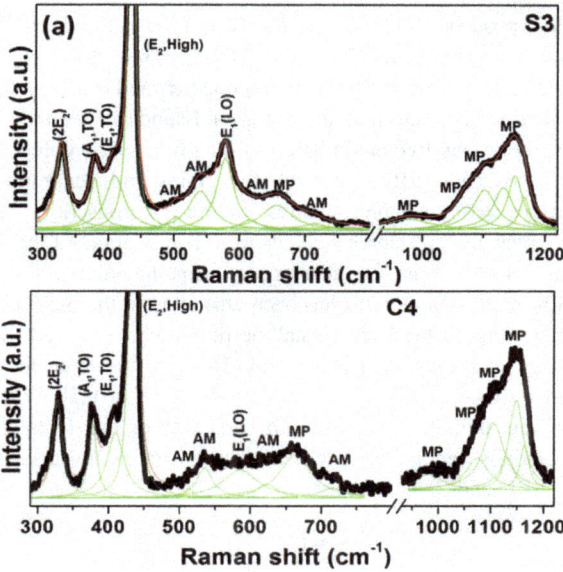

Figure 4.   The deconvoluted Raman spectrum of the samples synthesised through (a) the sol-gel route for Sample S3 and (b) the combustion route for Samples C4.

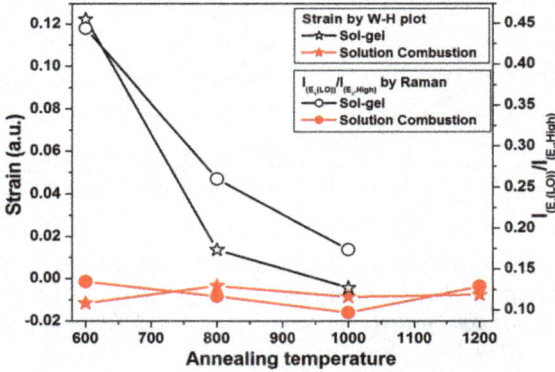

Figure 5.   Strain from W-H plot vs calcination temperature and $I_{(E1(LO))}/I_{(E2, High)}$ vs calcination temperature from Raman plots for ZnO powder synthesised through the sol-gel and the combustion route.

Fig. 6 depicts the positron lifetime spectra (PAL) best fitted with three lifetime components such as τ_1, τ_2, and τ_3 for S-series and C-series samples. The fitted parameters are shown in Table 2. The shortest lifetime component τ_1 and intensity I_1 corresponds to the positron annihilation at structural defects in grain boundaries. It may be noted that in a bulk material, τ_1 represents free annihilation of positrons in the defect-free region. But in nano materials, positrons diffuse through the grains and get trapped at the grain boundaries which are strong trapping centres for the positrons [33]. The intermediate lifetime, τ_2 with intensity I_2, represents positron annihilation at nano voids at the intersection of three or more grain boundaries (e.g. triple junctions). The longest lifetime component, τ_3 with intensity about 2% has been attributed to the pick-off annihilation of ortho positronium formed in the inter-crystalline region characterized by large volumes [34]. In ZnO, oxygen vacancy (V_o) is invisible in PAS measurement as the positron binding energy of V_o (0.04 eV) is much less than that of V_{Zn} (0.39 eV). Thus, one may observe only positron annihilation due to Zn vacancies or due to their clusters. With increase in calcinations temperature, τ_1, τ_2 and τ_3 gradually decrease in S-series corroborates with XRD and Raman results.

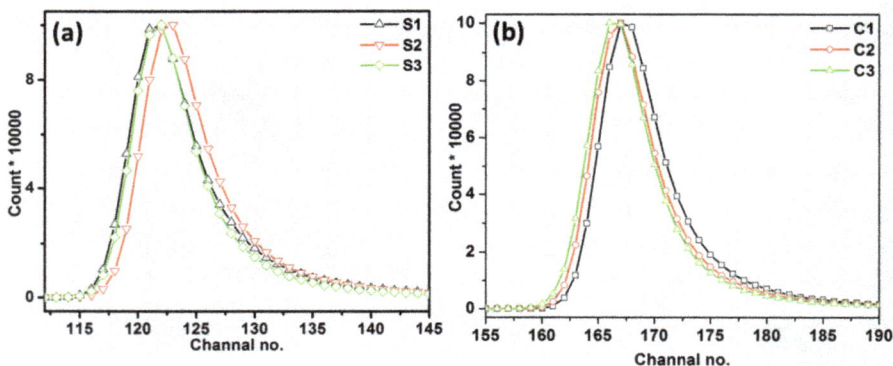

Figure 6.   Positron annihilation lifetime spectrum of ZnO powder synthesised through (a) the sol-gel route and calcined at different temperatures (S1 = 600, S2 = 800, S3 = 1000°C) and (b) the combustion route and calcined at different temperatures (C1 = 600, C2 = 800, C3 = 1000°C

Table 2. Positron annihilation lifetimes (τ) and corresponding intensities (I) of ZnO synthesized through the sol-gel and the combustion method, calcined at different temperatures.

ZnOcalci. (in °C)	Positron Lifetime (ps)			Defect Relative Intensities		
	τ_1	τ_2	τ_3	I_1	I_2	I_3
Sol-gel method						
S1	170.9±2.9	393.8±6.0	5661.8±1779.4	62.8±1.5	37.08±1.51	0.07±.02
S2	154..9±7.9	312.7±24.1	754.8±165.9	54.41±6.7	43.22±2.3	2.35±1.67
S3	142.6±13.7	261.8±46.2	534.9±120.5	52.6±5.6	42.05±4.36	5.35±1.03
Combustion method						
C1	137.1±9.0	275.6±13.2	840.2±104.8	43.13±6.2	54.53±5.6	2.34±.73
C2	143.4±12.6	259.7±32.7	617.7±111.1	52.15±13.8	44.04±11.79	3.82±2.35
C3	142.5±6.8	277.5±26.6	787.9±153.4	65.83±7.6	32.34±6.77	1.83±.96

τ_1, τ_2, and τ_3 are Positron Life Time with Relative Intensities I_1, I_2, and I_3

Photoluminescence spectra obtained at room temperature for S-series and C-series samples are shown in Figure 7 a & b, respectively. Usually, in ZnO, the luminescence spectrum shows near band edge (NBE) in UV region and defect band emission (DBE) in the visible range. While the former one arises due to recombination of free excitons and the latter is due to intrinsic defects [35,36]. In the present case, it is clear from Fig. 7(a) for S-series samples that exciting with 355 nm, a sharp, intense NBE emission is observed without showing DBE peak. For the C-series samples, there is a clear NBE emission at 392.6 nm and DBE in the range of 440 nm to 800 nm. The NBE peak is very sharp and intense whereas the DBE peak is broad. A slight decrease in FWHM of NBE peak with calcination temperature is due to higher crystalinity. However, the intensity of the broad DBE band for C-series samples increases with increase in calcination temperature, which is surprising. The intensity ratio of I_{DBE}/I_{NBE} is found to be 0.04, 0.12, 0.25 and 0.56 for C1, C2, C3 and C4, respectively. The deconvolution of C4 sample shown as inset of Fig. 8 demonstrates three sub bands centered at 504/2.46, 548/2.26 and 603/2.1 nm/eV. It is reported that V_{Zn} and V_o with their clusters are found to be responsible for DBE peak centered at 504/2.46 and 548/2.26 nm/eV, respectively [5, 37]. Mostly, the band about at 603/2.1 nm/eV is reported for oxygen interstitial (O_i) [36, 5].

We have observed red shifting of DBE emission with increasing the calcination temperature (Fig. 8).

The shifting and increase in intensity of DBE emission seems to be due to increase in the intensity of band at 603/2.1 nm/eV with annealing temperature.

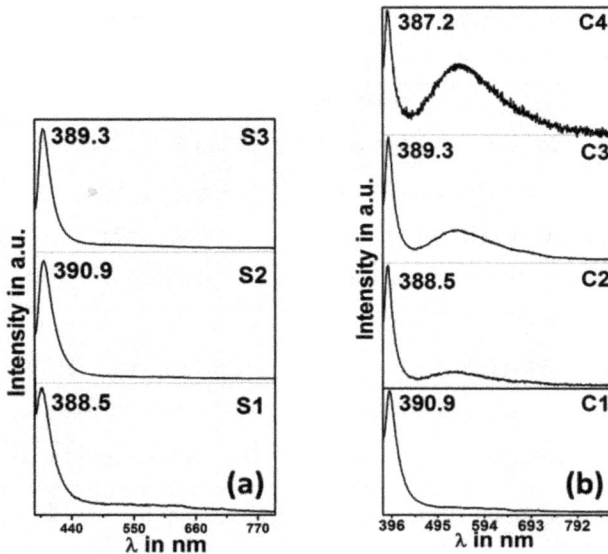

Figure 7. *The PL spectrum obtained by using the excitation wavelength 355 nm for the ZnO powder synthesised through (a) the sol-gel route and calcined at different temperatures (S1 = 600, S2 = 800, S3 = 1000 °C) and (b) the combustion route and calcined at different temperatures (C1 = 600, C2 = 800, C3 = 1000, C4 = 1200 °C).*

The increase in intensity of above DBE peak could be due to higher concentration of O_i defects which is produced due to annealing at higher temperature in air [36, 5, 38, 39]. However, the difference of DBE emission in-between sol-gel and combustion synthesized ZnO is surprising. While Raman spectra indicate decrease in native defects in both series samples, PL study demonstrate the presence of defects in C-series samples. The unusual behaviour could only be assigned to presence of different kinds of defects responsible for Raman anomaly and DBE emission. Ilyas et al. have observed the excess Zn_i under Zn-rich conditions and with increasing annealing temperature in O-rich environment Oi increases in ZnO [40]. Thus we confirm that while Zn_i is found to be responsible for Raman anomaly at low calcination temperature, O_i is found to be responsible for increase in DBE emission with increase in calcination temperature. The presence of O_i in C-series could be due to the combustion reaction between urea and nitrate precursor which generates elevated temperature that provides activation energy to create O_i inside the ZnO. Increasing the calcination temperature, O_i thus increases in C-series sample and show increase in intensity of DBE peak.

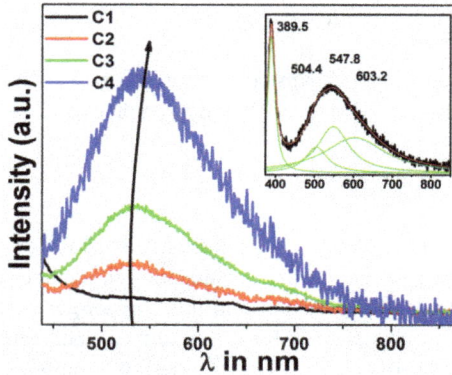

Figure 8. The normalised defect related deep band emission for ZnO samples synthesized through the combustion route (C1 = 600, C2 = 800, C3 = 1000, C4 = 1200 ^{o}C). Inset image shows the deconvoluted PL spectra of the sample C4.

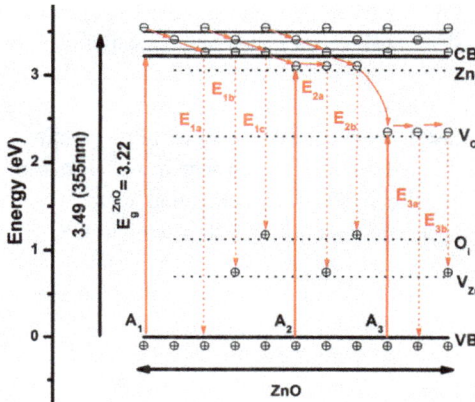

Figure 9. The proposed energy band diagram of the electron transfer in ZnO after excitation

Based on the Raman, PAS and photo luminescence study, we propose a tentative model to illustrate the mechanism of emission processes observed in ZnO (Fig.9). In the band diagram, we have specified the energy level of V_{Zn}, V_o, Zn_i and O_i corresponding to ZnO. Exciting with photons of energy 3.49eV (355nm), electrons may show band to band excitation and the formation of excitons. The relative absorption processes of electrons are represented by A1, A2 and A3 in the band diagram. Excited electrons are trapped by the defect states of Zn_i, V_o, O_i and V_{Zn} energy levels in the band gap of ZnO. The relative emission processes are E_{1a}, E_{1b}, E_{1c}, E_{2a}, E_{2b} and E_{3a}, E_{3b}, as shown in Fig. 9. We conclude that while strain decreases with increase in calcination temperature observation of high intense DBE peak in C-series sample can be applied as good luminescent material in visible range.

Conclusions

We investigated the intrinsic defects using various spectroscopic techniques such as RAMAN, Photoluminescence and Positron annihilation in pristine as well as in calcined ZnO particles synthesized by the combustion and the sol-gel techniques. From XRD patterns, pure wurtzite structure for both pristine and calcined samples were observed. However, few additional Raman modes indicated the presence of defects and/or impurities due to the breakdown of the translational symmetry of the lattice. $E_1(LO)$ mode ascribed to the presence of defects such as interstitial zinc which decreased with increase in calcination temperature irrespective of synthesis techniques. Positron life time spectra showed the decrease in defects with increase in calcination temperature in both samples. Williamson-Hall analysis indicated tensile and compressive strain respectively in samples made by sol-gel and combustion techniques. While tensile strain decreased in the former sample, an unusual compressive strain was found to be constant in the latter sample one with increase in calcination temperature. Photoluminescence spectra however demonstrated defects related broad emission which was absent in the former sample; in the latter, it increased with increase in calcination temperature without showing any strain. The increase in defects induced emission without change in strain in the lattice was discussed. Photoluminescence spectra however demonstrated defects related broad emission which was absent in the former sample; in the latter, it increased with increase in calcination temperature without showing any strain. Increase in defects induced emission without change in strain in the lattice was discussed.

Acknowledgements

UGC-DAE Consortium for Scientific Research, Kolkata Centre, is acknowledged for providing the PAS facilities.

References

[1] S. J. Pearton, D. P. Norton, K. Ip, Y. W. Heo, T. Steiner, Recent progress in processing and properties of ZnO, Superlattices Microstruct 34 (2003) (1–2) 3–32.

[2] H. Morkoç, Ü. Özgür, Zinc oxide: fundamentals, materials and device technology, John Wiley & Sons (2008).

[3] L. Schmidt-Mende , J. L. MacManus-Driscoll, ZnO – nanostructures, defects, and devices, Materials today 10, (2007) 40–48. http://dx.doi.org/10.1016/S1369-7021(07)70078-0

[4] Ü. Özgür, Y. I. Alivov, C. Liu, A. Teke, M. A. Reshchikov, S. Doğan,V. Avrutin, S. J. Cho, H. Morkoçd, A comprehensive review of ZnO materials and devices, J. Appl. Phys. 98 (2005) 041301. http://dx.doi.org/10.1063/1.1992666

[5] C. H. Ahn, Y. Y. Kim, D. C. Kim, S. K. Mohanta, H. K. Cho, A comparative analysis of deep level emission in ZnO layers deposited by various methods, J. Appl. Phys. 105 (2009) 013502-013502-5. http://dx.doi.org/10.1063/1.3054175

[6] P.S. Xu, Y.M. Sun, C.S. Shi, F.Q. Xu, H.B. Pan, The electronic structure and spectral properties of ZnO and its defects, Nucl. Instrum. Methods Phys. Res. Sect. B. 199 (2003) 286. http://dx.doi.org/10.1016/S0168-583X(02)01425-8

[7] S.B. Zhang, S.H. Wei, A. Zunger, Intrinsic n-type versus p-type doping asymmetry and the defect physics of ZnO, Phys. Rev. B. 63 (2001) 075205. http://dx.doi.org/10.1103/PhysRevB.63.075205

[8] P. Jiang, J. J. Zhou, H. F. Fang, C. Y. Wang, Z. L. Wang, S. S. Xie, Hierarchical shelled ZnO structures made of bunched nanowire arrays, Adv. Funct. Mater. 17 (2007) 1303-1310. http://dx.doi.org/10.1002/adfm.200600390

[9] V.P. Singh, D. Das and Chandana Rath, Studies on intrinsic defects related to Zn vacancy in ZnO nano particles, Materials Research Bulletin 48 (2013) 682. http://dx.doi.org/10.1016/j.materresbull.2012.11.026

[10] J. Qiu, Z. Jin, Z. Liu, X. Liu, G. Liu, W. Wu, X. Zhang, X. Gao, Fabrication of TiO₂ nanotube film by well-aligned ZnO nano rod array film and sol-gel process, Thin Solid Films, 515 (2007) 2897-2902. http://dx.doi.org/10.1016/j.tsf.2006.08.023

[11] A. Dev, S. K. Panda, S. Kar, S. Chakrabarti, S. Chaudhuri, Surfactant-assisted route to synthesize well-aligned ZnOnanorod arrays on sol-gel-derived ZnO thin films, J. Phys. Chem. B. 110 (2006) 14266-14272. http://dx.doi.org/10.1021/jp0627291

[12]   C. Kashinath, S.T. Patila, Arunab, T. Mimani, Combustion synthesis: an update, Current Opinion in Solid State and Materials Science 6 (2002) 507. http://dx.doi.org/10.1016/S1359-0286(02)00123-7

[13]   A. Sugunan, H. C. Warad, M. Boman, J. Dutta, Zinc oxide nano wires in chemical bath on seeded substrates: role of hexamine, J. Sol-Gel Sci. Technol. 39 (2006) 49-56. http://dx.doi.org/10.1007/s10971-006-6969-y

[14]   L. Znaidi, Sol-gel-deposited ZnO thin films, Mater. Sci. Eng. B. 174 (2010) 18-30. http://dx.doi.org/10.1016/j.mseb.2010.07.001

[15]   S. Roy, W. Sigmund, F. Aldinger, Nano structured yttria powders via gel combustion, Mater. Res. 14 (1999) 1524-1531. http://dx.doi.org/10.1557/JMR.1999.0204

[16]   T. Mimani, J. Alloys Compd. 315 (2001) 123-128; T. Mimani, K. C. Patil, Solution combustion synthesis of nano scale oxides and their composites, Mater. Phys. Mech. 4 (2001) 134-137.

[17]   S. Bhaduri, S. B. Bhaduri, K. A. Prisbrey, Auto ignition synthesis of nano crystalline $MgAl_2O_4$ and related nano composites, J. Mater.Res. 14 (1999) 3571-3580. http://dx.doi.org/10.1557/JMR.1999.0470

[18]   V. C. Sousa, A. M. Segadaes, M. R. Morelli, R. Kiminami, Combustion synthesized ZnO powders for varistor ceramics, J. Inorg. Mater. 1 (1999) 235-241. http://dx.doi.org/10.1016/S1466-6049(99)00036-7

[19]   P. Kirkegaard, N.J. Pedersen, M. Eldrup, RISØ-M-2740. PATFIT-88: A Data-processing System for Positron Annihilation Spectra on Mainframe and Personal Computers, Risø National Laboratory, DK-4000 Roskilde, Denmark, 1989, Information on http://www.risoe.dk/rispubl/reports/ris-m-2740.pdf.

[20]   G.K. Williamson, W.H. Hall, X-ray line broadening from filed aluminium and wolfram, Acta Metallurgica 1 (1953) 22-31. http://dx.doi.org/10.1016/0001-6160(53)90006-6

[21]   S. Dutta, S. Chattopadhyay, A. Sarkar, M. Chakrabarti, D. Sanyal, D. Jana, Role of defects in tailoring structural, electrical and optical properties of ZnO, Progress in Materials Science 54 (2009) 89-136. http://dx.doi.org/10.1016/j.pmatsci.2008.07.002

[22]   K. C. Barick, M. Aslam, V. P. Dravid, D. Bahadur, Controlled fabrication of oriented co-doped ZnO clustered nano assemblies, J. Colloid Interface Sci. 349 (2010) 19-26. http://dx.doi.org/10.1016/j.jcis.2010.05.036

[23]   V. Etacheri, R. Roshan, V. Kumar, Mg-doped ZnO nano particles for efficient
       sunlight-driven photo catalysis, ACS Appl. Mater. Interfaces 4 (2012) 2717-2725.
       http://dx.doi.org/10.1021/am300359h

[24]   R. Sendi, S. Mahmud, Stress Control in ZnO Nano particle-based Discs via High-
       Oxygen Thermal Annealing at Various Temperatures, J. Phys. Sci. 24 (2013) 1-15.

[25]   F. J. Manjon, B. Mari, J. Serrano, A. H. Romero, Silent Raman modes in zinc
       oxide and related nitrides, J. Appl. Phys. 97 (2005) 053516.
       http://dx.doi.org/10.1063/1.1856222

[26]   K. Samanta, P. Bhattacharya, R. S. Katiyar, W. Iwamoto, P. G. Pagliuso, C.
       Rettori, Raman scattering studies in dilute magnetic semiconductor $Zn_{1-x}Co_xO$,
       Phys. Rev. B. 73 (2006) 245213. http://dx.doi.org/10.1103/PhysRevB.73.245213

[27]   X. Xu, C. Cao, Hydrothermal synthesis of Co-doped ZnO flakes with room
       temperature ferromagnetism, J. Alloys Compd. 501 (2010) 265-268,

[28]   L. B. Duan, X. R. Zhao, J. M. Liu, T. Wang, G. H. Rao, Room-temperature
       ferromagnetism in lightly Cr-doped ZnO nano particles, Appl. Phys.A 99 (2010)
       679-683. http://dx.doi.org/10.1007/s00339-010-5590-7

[29]   D. Wang, Z. Q. Chen, D. D. Wang, N. Qi, J. Gong, C. Y. Cao, Z. Tang, Positron
       annihilation study of the interfacial defects in ZnO nanocrystals: Correlation with
       ferromagnetism, J. Appl. Phys. 107 (2010) 023524-023524-8.
       http://dx.doi.org/10.1063/1.3291134

[30]   S. K. Sharma, G. J. Exarhos, Raman spectroscopic investigation of ZnO and doped
       ZnO films, nanoparticles and bulk material at ambient and high pressures, Solid
       State Phenomena 55 (1997) 32-37.
       http://dx.doi.org/10.4028/www.scientific.net/SSP.55.32

[31]   V.P. Singh, R.K.Singh, D.Das, ChandanaRath, Defects in $Zn_{1-x-y}Co_xMg_yO$
       nanoparticles: Probed by XRD, RAMAN and PAS techniques, Materials Science
       in Semiconductor Processing 16 (2013) 659–666.
       http://dx.doi.org/10.1016/j.mssp.2012.12.006

[32]   J. Serrano, A. H. Romero, F. J. Manjon, R. Lauck, M. Cardona, A. Rubio, Pressure
       dependence of the lattice dynamics of ZnO: An ab initio approach, Phys. Rev. B
       69 (2004) 094306. http://dx.doi.org/10.1103/PhysRevB.69.094306

[33]   A. K. Mishra, S. K. Chaudhuri, S. Mukherjee, A. Priyam, A. Saha, D. Das,
       Characterization of defects in ZnO nanocrystals: Photoluminescence and positron

annihilation spectroscopic studies, J. Appl. Phys. 102 (2007) 103514. http://dx.doi.org/10.1063/1.2817598

[34] D. C. Look, D. C. Reynolds, J. R. Sizelove, R. L. Jones, C. W. Litton, G. Cantwell, W. C. Harsch, Electrical properties of bulk ZnO, Solid State Commun. 105 (1998) 399-401. http://dx.doi.org/10.1016/S0038-1098(97)10145-4

[35] J. Liu, Y. Zhao, Y. J. Jiang, C. M. Lee, Y. L. Liu, G. G. Siu, Identification of zinc and oxygen vacancy states in non polar ZnO single crystal using polarized photoluminescence, Appl. Phys. Lett. 97 (2010) 231907. http://dx.doi.org/10.1063/1.3525714

[36] N. H. Alvi, W. ul Hassan, B. Farooq, O. Nur, M. Willander, Influence of different growth environments on the luminescence properties of ZnO nanorods grown by the vapor-liquid-solid (VLS) method, Mater. Lett. 106 (2013) 158-163. http://dx.doi.org/10.1016/j.matlet.2013.04.074

[37] L. Ke, S. C. Lai, J. D. Ye, V. L. Kaixin, S. J. Chua, Point defects analysis of zinc oxide thin films annealed at different temperatures with photoluminescence, Hall mobility, and low frequency noise, J. Appl. Phys. 108 (2010) 084502. http://dx.doi.org/10.1063/1.3494046

[38] A. Janotti, C. G. Van de Walle, Fundamentals of zinc oxide as a semiconductor, Rep. Prog. Phys. 72 (2009) 126501. http://dx.doi.org/10.1088/0034-4885/72/12/126501

[39] R. Sendi, S. Mahmud, Stress Control in ZnO Nano particle-based Discs via High-Oxygen Thermal Annealing at Various Temperatures, J. Phys. Sci. 24 (2013) 1-15.

[40] U. Ilyas, R. S. Rawat, T. L. Tan, P. Lee, R. Chen, H. D. Sun, L. Fengji, S. Zhang, Oxygen rich p-type ZnO thin films using wet chemical route with enhanced carrier concentration by temperature-dependent tuning of acceptor defects, J. Appl. Phys. 110 (2011) 093522. http://dx.doi.org/10.1063/1.3660284

CHAPTER 6

Bonding in $La_{0.9}Zn_{0.1}FeO_3$ multiferroic material

G.Gowri*, R.Saravanan , R.Pradeepa, M.Raja Rajeswari, K.Abirami

Research centre and Post Graduate Department of Physics, The Madura College,Madurai-625011, Tamil Nadu, India

*gowrikanna01@gmail.com; saragow@gmail.com

Abstract

In this work, zinc doped LaFeO3 multiferroic ($La_{0.9}Zn_{0.1}FeO_3$) has been prepared by the chemical co-precipitation method. The prepared sample has been characterized using powder X-ray diffraction, transmission electron microscope, UV-Visible spectrometer and vibration sample magnetometer respectively. The structural analysis has been done on the powder X-ray diffraction data of the sample using the powder profile refinement technique and the results obtained from the refinement process have been used to analyse the electron density distribution and also bonding nature between the neighbouring atoms in the unit cell of the prepared sample using the Maximum Entropy Method (MEM). The average particle size is determined using TEM images. The optical band gap energy is estimated using UV-Visible absorption spectrum. The magnetic parameters are extracted from the hysteresis loop recorded using a vibration sample magnetometer (VSM).

Keywords

LaFeO3, X-Ray Diffraction, Maximum Entropy Method, Transmission Electron Microscope, Coercivity

Contents

1.   Introduction

In recent decades, many researchers have concentrated their studies on doped $LaFeO_3$ due to their wide applications in numerous advanced technologies such as non-volatile magnetic memory devices, solid oxide fuel cells and ultrasensitive magnetic read heads of modern hard disk drives etc. [1-3]. $LaFeO_3$ is of ABO_3 perovskite structure with orthorhombic phase, which is derived from the standard cubic structure through the distortion of the BO_6 octahedra [4-6]. In ABO_3 perovskite materials, the smaller cations B are at the centre of the octahedron of oxygen anions and the large cations A are at the corners of the unit cell [7,8]. Since $LaFeO_3$ has interesting properties like, wide-gapped antiferromagnetic insulator with high Néel temperature ($T_N \sim 740°C$) [9], coexistence of coupled ferroelectric and antiferromagnetic ordering etc., it is considered as a significant material among the rare earth orthoferrite series [10]. Though, the pure and doped lanthanum ferric oxide can be prepared by various process like microwave synthesis [11-13], solid state reaction method [14], hydrothermal method [15-17], and the chemical co-precipitation method [18] etc., the chemical co-precipitation method is the most attractive one because, it is the most cost effective method. Research on pure and doped lanthanum orthoferrite reveals that their structural properties strongly depend on the synthesis and processing of its precursor powders. Research also reveals that the properties of $LaFeO_3$ can be altered or modified precisely by the choice of suitable doping element at the La^{3+} or at the Fe^{3+} site in that system. It has been already reported that in the $LaFeO_3$ multiferroic system, Zn^{2+} doping at the La^{3+} site alters its magnetic behaviour. The magnetization of $La_{1-x}Zn_xFeO_3$ (x = 0.1, 0.3) multiferroics were found to increase with the increase in dopant concentration [18]. It is well known that electronic structure of materials and the bonding are important parts in materials characterization. This has enthused the authors to synthesis $La_{0.9}Zn_{0.1}FeO_3$ multiferroic and to analyse the charge density distribution and also bonding nature of atoms in their unit cell that are responsible for the optical and magnetic behaviour of the material.

2. Experimental Procedure

2.1 Sample Preparation

Zinc doped lanthanum orthoferrite ($La_{0.9}Zn_{0.1}FeO_3$) used in this work was prepared by chemical co-precipitation method using high purity oxides La_2O_3 (99.99 %, Alfa Aesar), ZnO (99.99 %, Alfa Aesar) and Fe_2O_3 (99.99 %, Alfa Aesar) as starting materials. These oxides in the desired stoichiometric proportions were weighted and each oxide was taken in a beaker and required amount of distilled water was added, until the metal powder becomes slurry. To get a clear salt solution from this slurry, concentrated nitric acid was added drop-wise in each oxide and finally these three solutions were taken in a single beaker. Then the solution was ultrasonicated for about 2h. For the complete precipitation of metal oxides from this solution under ultrasonication, 0.4 N NaOH solution was added until its pH reaches ~ 9. The co-precipitated particles were heated at 80°C with rigorous stirring by a magnetic stirrer for 2h. The washing of the obtained co- precipitate was carried out with the help of water and was separated using vacuum filter. Finally, the co-precipitate was dried at 100 °C for 16 h. The final product was pre-calcined in a furnace at 400 °C for 3 h. The precursor was made into circular pellets of 12 mm in diameter and 1 mm thickness and then sintered at temperature of 900 °C for 2 h, to obtain the $La_{0.9}Zn_{0.1}FeO_3$ multiferroic.

2.2 Results and Discussion

2.2.1 X-Ray Profile Analysis

The powder XRD data was obtained for Zn doped lanthanum orthoferrite ($La_{0.9}Zn_{0.1}FeO_3$) using X-ray diffractometer, Bruker AXS D8 Advance, with CuK_α monochromatic beam at sophisticated analytical instrument facility (SAIF), Cochin university, Cochin, India, in the 2θ range of 5°-120° with the step size of 0.02° in 2θ. Fig. 1 shows the XRD pattern of the prepared multiferroic.

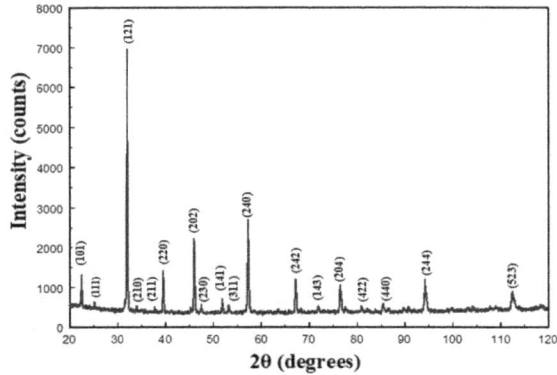

Figure 1. Observed powder XRD profile of $La_{0.9}Zn_{0.1}FeO_3$ multiferroic.

All the observed X-ray peaks for the prepared orthoferrite were identified and matched with the $LaFeO_3$ phase of the standard pattern of the Joint Committee for Powder Diffraction Standards (JCPDS) XRD data set reported in the file (No.37-1493). The XRD result indicates that the product is a perovskite oxide with orthorhombic system with the space group of Pnma. No additional phase has been identified in the XRD data. The structural parameters of prepared multiferroic were obtained by subjecting the X-ray diffraction data to the well-known Rietveld refinement technique [19], which is employed in the software JANA2006 [20]. The profile refinement [21] is the standard tool which was devised by Hugo Rietveld [19] for use in the characterization of crystalline materials. In this technique, cell parameters, structural parameters, peak shift, preferred orientation parameters, background profile shape are used to minimize the difference between the theoretically modelled profile and the observed one.

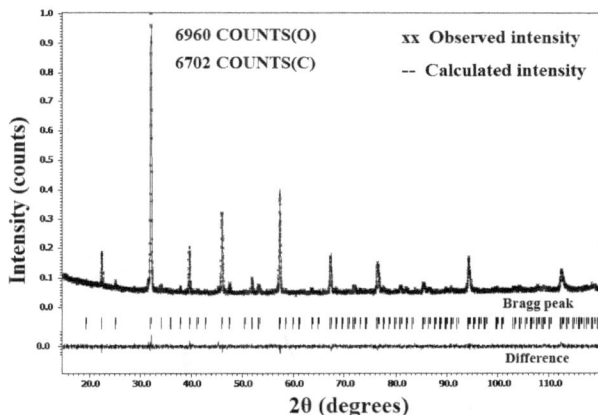

Figure 2. Rietveld refined profile of $La_{0.9}Zn_{0.1}FeO_3$ mutiferroic.

With the powder XRD data, the refinement was done taking into account orthorhombic structure for $La_{0.9}Zn_{0.1}FeO_3$ multiferroic having space group of Pnma with 4 molecules per unit cell. The refined profile for the prepared orthoferrite is shown in Fig. 2 and the refined structural parameters are given in Table 1. The refined positional coordinates are given in Table 2.

Table 1. Structural parameters of $La_{0.9}Zn_{0.1}FeO_3$ multiferroic

Parameter	$LaFeO_3$
a (Å)	5.5637(2)
b (Å)	7.8515(3)
c (Å)	5.5536(2)
$\alpha = \beta = \gamma$ (°)	90
Cell Volume, V ($Å^3$)	242.6022(12)
Density, ρ (gm/ cc)	6.4428(3)
Observed reliability factor, R_{obs} (%)	2.94
Profile reliability factor, Rp (%)	3.92
GOF	1.07
Number of electrons in the unit cell	417

67

Table 2. Refined atomic positional coordinates of $La_{0.9}Zn_{0.1}FeO_3$ multiferroic

Atom	LaFeO₃		
	x	y	z
La/Al	0.0279	0.25	0.9939
Fe	0.5	0	0
O₁	0.4980	0.25	0.0710
O₂	0.2645	0.0417	0.7101

2.2.2 Charge Density Analysis

For the chosen system, to study the charge density distribution accurately and hence to understand the nature of chemical bonds between the neighbouring atoms in the unit cell, the Maximum Entropy Method (MEM) was used. Maximum Entropy Method (MEM) which is a statistical approach proposed by Collins [22], is a powerful tool to study the electronic structure of the materials. The refined structural parameters evolved from the Rietveld refinement technique were used for the MEM procedure to construct the charge density distributions around the atoms in the unit cell. In the Maximum Entropy Method, the structure factors were calculated iteratively while increasing the entropy in the charge density in every cycle until the convergence criterion becomes 1 under minimum possible iterations. The charge density of the sample was estimated using the software PRIMA (Practice of Iterative MEM Analysis) [23], which employs the Maximum Entropy Method technique and the resultant charge density distribution has been mapped and visualized using the visualization software VESTA (Visualization for Electronic and STructural Analysis) [24]. The MEM refinement in this work was carried out by dividing the unit cell into 48 x 72 x 48 pixels. The MEM refined parameters are given in Table 3.

Table 3. Parameters from MEM refinements of $La_{0.9}Zn_{0.1}FeO_3$ multiferroic

Parameter	LaFeO$_3$
Number of pixels in the unit cell (48x72x48)	165888
Number of electrons in the unit cell	417
Lagrangian parameter, λ	0.012805
Number of refinement cycles	1562
Reliability factor, R_{MEM} (%)	1.59
Weighted reliability factor, wR_{MEM} (%)	1.96

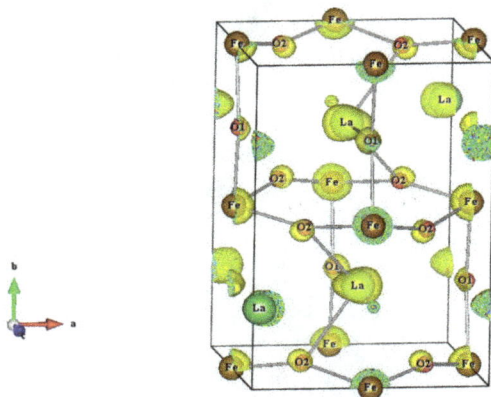

Figure 3.   3D electron density distribution in the unit cell (isosurface level = 9 e/ $Å^3$) for $La_{0.9}Zn_{0.1}FeO_3$ multiferroic.

Fig. 3 shows the three dimensional electron density distribution of $La_{0.9}Zn_{0.1}FeO_3$ multiferroic. The 3D electron density distribution clearly gives the position of the La, Fe and O atoms in the unit cell and confirms the orthorhombic phase of $La_{0.9}Zn_{0.1}FeO_3$ multiferroic. The shaded region enclosed by the electron cloud illustrates an atom. The bond length obtained from MEM is given in Table 4.

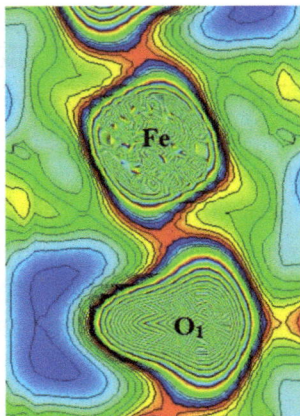

Figure 4. 2D electron density distribution in the unit cell showing Fe–O_1 bond in the (1 0 0) plane of $La_{0.9}Zn_{0.1}FeO_3$ multiferroic (Contour range is from 0 to 1.4 e/ $Å^3$ and contour interval is 0.1 e/ $Å^3$).

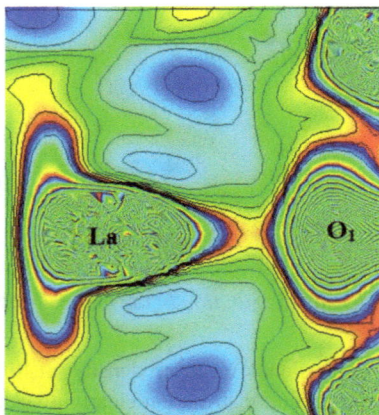

Figure 5. 2D electron density distribution in the unit cell showing La–O_1 bond in the (2 0 0) plane of $La_{0.9}Zn_{0.1}FeO_3$ multiferroic (Contour range is from 0 to 1.4 e/ $Å^3$ and contour interval is 0.12 e/ $Å^3$).

Fig. 4 depicts the two dimensional electron density distribution showing Fe–O_1 bond, in the (100) miller plane. Similarly, Fig.5 depicts the two dimensional electron density distribution showing La–O_1 bond in the (200) miller plane. From the 2D electron density map, it is clearly seen that the contour lines are elongated from Fe to O_1 atom and also from La to O_1 atom, signifying the non spherical distribution of charges that arises due to the sharing of electrons in the bonding region, which characterizes the bond Fe–O_1 and La–O_1 as covalent.

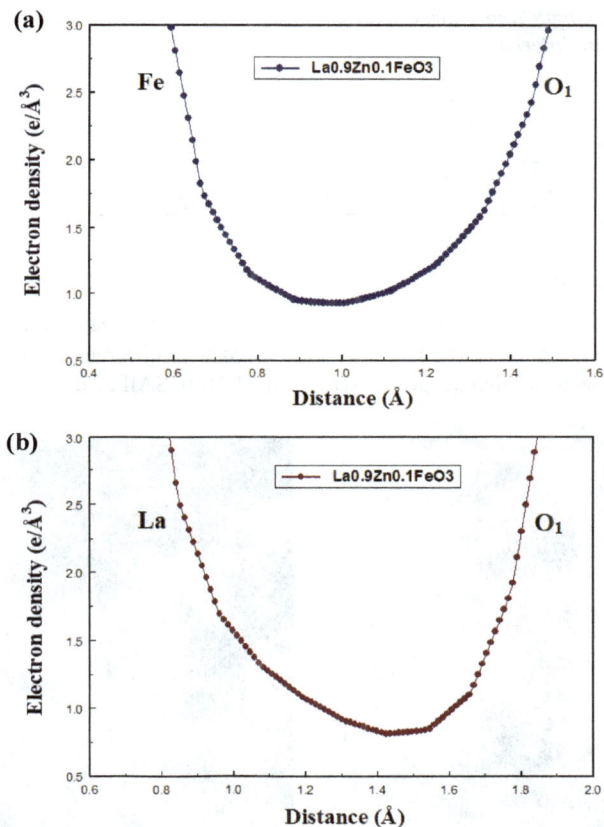

Figure 6. 1D profile showing variation of electron density along (a) Fe–O_1 bond, (b) La–O_1 bond.

In order to have a better understanding of the electron density distribution and the bonding nature, the profile of one dimensional electron density distribution is drawn for the Fe–O_1 and La–O_1 bond as shown in Fig. 6. The value of the charge density at the bond critical point along the Fe–O_1 bond and the La–O1 bond is estimated and is given in

Table 4. In the prepared multiferroic, for both the bonds, the value of the charge density at the bond critical point of Fe–O_1 and La–O_1 bond is high, so that the bonding can be declared as highly covalent.

Table 4. Bond length and charge density at the bond critical point between atoms in La$_{0.9}$Zn$_{0.1}$FeO$_3$ multiferroic

Bond	Distance (Å)	Charge density (e/Å^3)	Bond length (Å)
Fe-O_1	0.9960	0.9221	2.0008
La-O_1	1.4241	0.8119	2.4398

2.2.3 TEM Analysis

To find the particle size and to study the surface morphology of the synthesized La$_{0.9}$Zn$_{0.1}$FeO$_3$ multiferroic, TEM images were taken with different magnification using Transmission Electron Microscope (TEM) Jeol/JEM 2100, SAIF, Cochin, India.

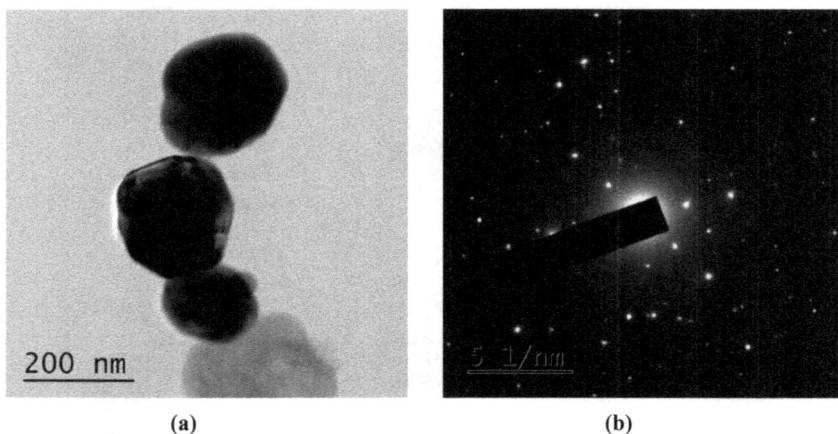

(a)   (b)

Figure 7. (a) TEM image (b) SAED pattern of La$_{0.9}$Zn$_{0.1}$FeO$_3$ multiferroic.

Fig. 7 illustates the TEM micrograph of La$_{0.9}$Zn$_{0.1}$FeO$_3$ multiferroic along with the corresponding selected area electron diffraction (SAED) pattern. The selected area electron diffraction pattern indicates the polycrystalline nature of the sample. From the TEM image, it can be seen that the particles are almost rounded hexagonal in shape and they seem to be regularly agglomerated. An overview of the TEM images of the prepared multiferroic shows that the particles have a size distribution of approximately 226 nm. The crystallite size of the prepared sample is also determined using Debye's Scherrer's formula $t = k\lambda/\beta \cos\theta$, where t is the crystallite size which is different from particle size, k is a constant usually of 0.9, λ the wavelength of X-rays (1.54056 A°) for CuK_α radiation, β is the full width at half maximum in radian for the prominent intensity peak, and θ is the Bragg angle of the reflection. The estimated crystallite size of the prepared multiferroic is found to be approximately 52 nm.

2.2.4 UV Analysis

For the prepared $La_{0.9}Zn_{0.1}FeO_3$ multiferroic, UV-visible absorption spectral analysis has been done. UV-visible absorption spectrum was recorded at SAIF, Cochin, India, using a UV-visible–NIR spectrometer, Varian Cary 5000. UV-visible optical absorption spectrum of the prepared $La_{0.9}Zn_{0.1}FeO_3$ multiferroic shows strong absorption peak in the UV-region at the wavelength approximately 389 nm.

The optical band gap for the prepared multiferroic is determined according to the Tauc equation $(\alpha h\nu) = A (h\nu - E_g)^n$ [25], where α is absorption coefficient, $h\nu$ is photon energy (E), A is a material constant, E_g is the energy band gap and n is 1/2, as $LaFeO_3$ system is a direct band gap material.

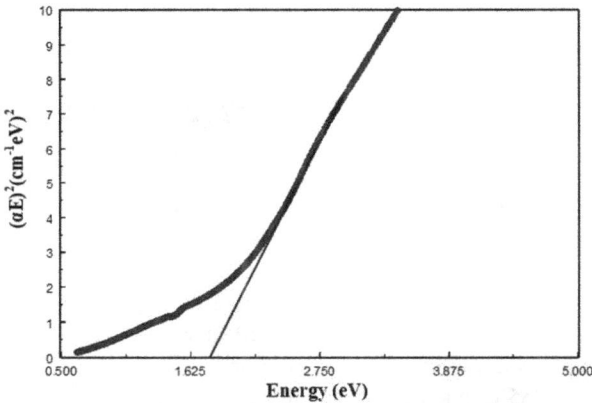

Figure 8. Plot of energy versus $(\alpha E)^2$ for $La_{0.9}Zn_{0.1}FeO_3$ multiferroic.

The plot of $(\alpha E)^2$ versus energy is drawn and is used to estimate the energy band gap of the $La_{0.9}Zn_{0.1}FeO_3$ multiferroic, where the extrapolation of the tangent of the curve to the zero value of $(\alpha E)^2$, gives the value of E_g as 1.79 eV as shown in Fig.8. It has already been reported that the optical energy band gap of pure $LaFeO_3$ ceramics was found to be vary from 2.12 eV to 2.51 eV [26]. The estimated energy band gap value of the prepared sample shows that Zn doping in $LaFeO_3$ system decreases its energy band gap value.

2.2.5 Magnetic Behaviour Analysis

To analyse the magnetic behaviour of the prepared $La_{0.9}Zn_{0.1}FeO_3$ multiferroic, it is subjected to the magnetic field and the magnetic hysteresis loop has been recorded at room temperature and hence the magnetic parameters are also measured using a vibration sample magnetometer (Lakeshore VSM) at sophisticated analytical instrument facility (SAIF), Indian Institute of Technology of Madras, Chennai, India.

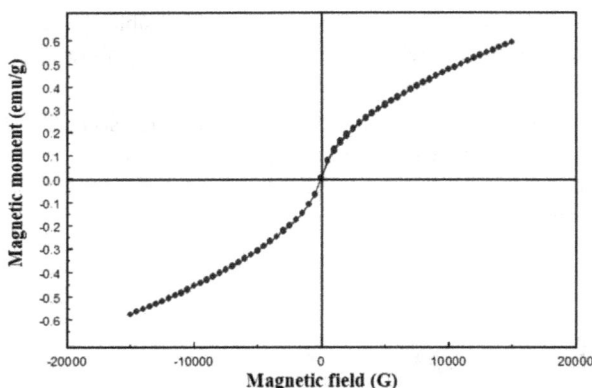

Figure 9. Magnetic hysteresis (M vs H) loop of $La_{0.9}Zn_{0.1}FeO_3$ multiferroic.

Fig. 9 shows the magnetic hysteresis loop of the prepared multiferroic recorded at room temperature. The values of coercivity, maximum magnetization and remanent magnetization extracted from the hysteresis loop are 2.9828 G, 0.584 emu/g, 0.0002 emu/g respectively. The sigmoidal shape of magnetization curve without any loop with very low coercivity and almost zero remanence reflects the superparamagnetic behaviour of the sample. It has been reported earlier that the maximum magnetization value of pure $LaFeO_3$ system is 0.485 emu/g and for the $La_{0.9}Zn_{0.1}FeO_3$ multiferroic the magnetization value is high as 0.659 emu/g [18]. The maximum magnetization value measured for the sample prepared in this sudy is found to be 0.584 emu/g, which agrees well with the reported value. Thus, the substitution of Zn^{2+} ions in $LaFeO_3$ reduces the magneto crystalline anisotropy due to which magnetization increases and at the same time the coercivity decreases. Since, $La_{0.9}Zn_{0.1}FeO_3$ multiferroic has low coercivity and moderately high value of magnetization; it can be used as a promising candidate in switching and other multifunctional devices.

Conclusions

$La_{0.9}Zn_{0.1}FeO_3$ multiferroic was successfully synthesized by the chemical co-precipitation method. The XRD analysis reveals that the prepared sample exhibits an orthorhombic phase of space group Pnma. The refined structure factors obtained from the Rietveld refinement method were used in the charge density analysis through MEM. The estimated value of the charge density at the bond critical point of $Fe-O_1$ and $La-O_1$ bond is found to be high as 0.9221 and 0.8119 respectively, which is good evidence of the strength of the covalent bonding in $La_{0.9}Zn_{0.1}FeO_3$ multiferroic. TEM images show that the particles are of almost rounded hexagonal in shape. The UV-visible spectrum of the sample was used to estimate its energy band gap value and it is found that Zn doping in the $LaFeO_3$ system decreases its energy band gap significantly. Further, the magnetic measurements show that Zn doping in $LaFeO_3$ increases its magnetization and decreases its coercivity.

Acknowledgement

The authorities of The Madura College are gratefully acknowledged for their constant encouragement of the research activities of the authors.

References

[1]   T. Hibino, S. Wang, S. Kakimoto, M. Sano, One-chamber solid oxide fuel cell constructed from a YSZ electrolyte with a Ni anode and LSM cathode, Solid State Ionics. 127 (2000) 89–98. http://dx.doi.org/10.1016/S0167-2738(99)00253-2

[2]   A. Moser, C.T. Rettner, M.E. Best, E.E. Fullerton, D. Weller, M. Parker, M.F. Doerner, Writing and detecting bits at 100 Gbit/in^2 in longitudinal magnetic recording media, IEEE Trans. Magn. 36 (2000) 2137–2139. http://dx.doi.org/10.1109/20.908333

[3]   S.S.P. Parkin, K.P. Roche, M.G. Samant, P.M. Rice, R.B. Beyers, R.E. Scheuerlein, E.J. O'Sullivan, S.L. Brown, J. Bucchigano, D.W. Abraham, Yu Lu, M. Rooks, P.L. Trouilloud, R.A. Wanner, W.J. Gallagher, Giant Magneto-Resistance Devices, J. Appl. Phys. 85 (1999) 5828–5833. http://dx.doi.org/10.1063/1.369932

[4]   Y.-H. Lee, J.-M. Wu, Epitaxial growth of LaFeO$_3$ thin films by RF magnetron sputtering, J. Cryst. Growth. 263 (2004) 436-441. http://dx.doi.org/10.1016/j.jcrysgro.2003.12.007

[5]   P.M. Woodward, Octahedral tilting in perovskites: I Geometrical considerations, Acta Crystallogr. B53 (1997) 32-43. http://dx.doi.org/10.1107/S0108768196010713

[6]   P.M. Woodward, Octahedral tilting in perovskites: II Structure stabilizing forces, Acta Crystallogr. B53 (1997) 44-66. http://dx.doi.org/10.1107/S0108768196012050

[7]   J.-M. Liu, Q.C. Li, X.S. Gao, Y. Yang, X.H. Zhou, X.Y. Chen, et al, Order coupling in ferroelectromagnets as simulated by a Monte Carlo method, Phys. Rev. B, 66, (2002) 054416–054426. http://dx.doi.org/10.1103/PhysRevB.66.054416

[8]   N.A. Hill, Why are there so few magnetic ferroelectrics? J. Phys. Chem. B, 104 (2000) 6694–6709. http://dx.doi.org/10.1021/jp000114x

[9]   G.R. Hearne, M.P. Pasternak, Electronic structure and magnetic properties of LaFeO$_3$ at high pressure, Phys. Rev. B, 51, (1995) 11495–11. http://dx.doi.org/10.1103/PhysRevB.51.11495

[10]   S. Acharya, J. Mondal, S. Ghosh, S.K. Roy, P.K. Chakrabarti, Multiferroic behavior of Lanthanum orthoferrite (LaFeO$_3$), Mater. Lett. 64 (2010) 415-418. http://dx.doi.org/10.1016/j.matlet.2009.11.037

[11]   Tang P, Tong Y, Chen H, Cao F, Pan G, Microwave-assisted synthesis of nanoparticulate perovskite LaFeO$_3$ as a high active visible-light photo catalyst, Curr. Appl. Phys. 13, (2013) 340–343. http://dx.doi.org/10.1016/j.cap.2012.08.006

[12]   Farhadi S, Momeni Z, Taherimehr M, Rapid synthesis of perovskite-type LaFeO3 nanoparticles by microwave-assisted decomposition of bimetallic La[Fe(CN)6] 5H2O compound, J. Alloy Compd. 471 (2009) 15–18. http://dx.doi.org/10.1016/j.jallcom.2008.03.113

[13]   Ding JLX, Shu H, Xie J, Zhang H, Microwave-assisted synthesis of perovskite ReFeO3 (Re: La, Sm, Eu, Gd) photocatalyst, Mater. Sci. Eng. B. 171 (2010) 31–34. http://dx.doi.org/10.1016/j.mseb.2010.03.050

[14]   S. Acharya, P.K. Chakrabarti, Some interesting observations on the magnetic and electric properties of Al doped lanthanum orthoferrite (La$_{0.5}$Al$_{0.5}$FeO$_3$), Solid State Commun. 150, (2010) 1234-1237. http://dx.doi.org/10.1016/j.ssc.2010.04.006

[15]   Thirumalairajan S, Girija K, Ganesh I, Mangalaraj D, Viswanathan C, Balamurugan A, Ponpandian N, Controlled synthesis of perovskite LaFeO$_3$ microsphere composed of nanoparticles via self-assembly process and their associated photo catalytic activity, Chem. Eng. J. 209 (2012) 420–428. http://dx.doi.org/10.1016/j.cej.2012.08.012

[16]   Zheng W, Liu R, Peng D, Meng G, Hydrothermal synthesis of LaFeO3 under carbonate-containing medium , Mater. Lett. 43 (2000) 19–22. http://dx.doi.org/10.1016/S0167-577X(99)00223-2

[17]   Ji K, Dai H, Deng J, Song L, Xie S, Han W, Glucose-assisted hydrothermal preparation and catalytic performance of porous LaFeO$_3$ for toluene combustion, J. Solid State Chem. 199 (2013) 164–170. http://dx.doi.org/10.1016/j.jssc.2012.12.017

[18]   K..Mukhopadhyay, A.S.Mahapatra, P.K.Chakrabarti, Multiferroic behavior, enhanced magnetization and exchange bias effect of Zn substituted nanocrystalline LaFeO$_3$ (La$_{(1-x)}$Zn$_x$FeO$_3$, x=0.10, and 0.30), J. Magn. Magn. Mater. 329 (2013) 133–141. http://dx.doi.org/10.1016/j.jmmm.2012.09.063

[19]   H.M. Rietveld, A Profile Refinement Method for Nuclear and Magnetic Structures, J. Appl. Crystallogr. 2 (1969) 65-71. http://dx.doi.org/10.1107/S0021889869006558

[20]   Petříček, V., Dušek, & M., Palatinus, L. (2006). JANA 2006, the crystallographic computing system. Praha, Czech Republic, Academy of Sciences of the Czech Republic.

[21]   M. M. Wolfson, Introduction to X-ray Crystallography. Cambridge University Press, London, 1970.

[22]   D.M.Collins, Electron density images from imperfect data by iterative entropy maximization, Nature. 298 (1982) 49-51. http://dx.doi.org/10.1038/298049a0

[23]   F.Izumi, R.A.Dilanien, PRIMA, for the maximum entropy method advanced materials laboratory, Japan (2004).

[24]   K.Momma and F.Izumi, VESTA: a three-dimensional visualization system for electronic and structural analysis, J.Appl. Crystallogr. 41 (2008) 653. http://dx.doi.org/10.1107/S0021889808012016

[25]   J.Tauc, R.Grigorvici, A.Vancu, Optical properties and electronic structure of amorphous germanium, Physica status solidi 15, 627-637 (1966). http://dx.doi.org/10.1002/pssb.19660150224

[26]   Roberto Köferstein, Lothar Jäger, Stefan G. Ebbinghaus, Magnetic and optical investigations on LaFeO$_3$ powders with different particle sizes and corresponding ceramics, Solid State Ionics, 249-250 (2013) 1-5. http://dx.doi.org/10.1016/j.ssi.2013.07.001

CHAPTER 7

Magnetic phase transition in $Co_{0.5}Zn_{0.5}Fe_2O_4$ nanoparticles

Muhammad Misbah Muhammad Zulkimi[1], Ismayadi Ismail[1*],

[1]Materials Synthesis and Characterisation Laboratory, Institute of Advanced Technology (ITMA), Universiti Putra Malaysia,43400 Serdang, Selangor, Malaysia

*Email: kayzen@gmail.com

Abstract

The objective of this work is to expose and explain experimental microstructure property relationship with magnetic properties as they evolved with increasing sintering temperature. Mechanical alloying (MA) was used to prepare $Co_{0.5}Zn_{0.5}Fe_2O_4$ nanoparticles with sintering temperature from 600°C to 1350°C with an increment of 50°C. The phase changes, grain size and magnetic properties were measured. Magnetic parameter, namely, hysteresis loop was measured at room temperature for samples sintered from 600°C to 1350°C. The results showed that Cobalt zinc ferrite cannot be formed directly through milling alone, but heat treatment was necessary. After annealing the sample at 600°C, cobalt zinc ferrite phase was first obtained with an average grain size in the nanometric range (0.089μm). B-H hysteresis loops could be categorized into three distinct groups according to their shapes. The activation energy value increased with increasing grain size caused by the increasing temperature, and found to be 7.58 kJ/mol, 47.11 kJ/mol and 189.54 kJ/mol. The slopes of the activation energy corresponded to the three different groups for magnetic properties (B-H loops). Meanwhile, the resistivity could be classified into two parts based on their shape of the graph, and their values decreased with increasing temperature which confirmed that the sintered samples under investigation possessed an semiconducting behavior.

Keywords

Mechanical Alloying, Magnetic Parameters, Resistivity

Contents

1. Introduction

There is a growing interest in magnetic nanoparticles, due to their broad practical application in several important technological fields such as ferrofluid, magnetic drug delivery, and high-density information storage. Due to these, great attention has been given to the preparation and characterization of the metal oxide nanoparticles of spinel ferrite, MFe_2O_4 (M= Co, Mg, Mn, Ni, etc) [1].

Spinel ferrites have high electrical resistivity and low eddy current and dielectric losses and therefore, they are found to be very useful for technological application. The spinel ferrite has been fabricated by many techniques such as sol-gel [2,3], co-precipitation method etc. [4,5]. However, complex process, expensive precursors, and low production yields are common problems.

In the evolution work, Idza et al. [6] have reported that the $Ni_{0.3}Zn_{0.7}Fe_2O_4$ toroidal samples were prepared via high-energy ball milling and subsequent moulding; the samples with nanometer/submicron sized compacted powder were sintered from 600°C to 1400°C using 100°C increments [6]. An integrated analysis of phase, micro structural and hysteresis data would point to the existence of three distinct shape-differentiated groups of B–H hysteresis loops which belong to samples with weak, moderate and strong magnetism. Ismayadi et al. [7] have reported the microstructure and the magnetic properties and showed that the importance of grain-size threshold for the appearance of significant ordered magnetism (mainly ferromagnetism) was about ≥0.3 μm. They found that there were three stages of magnetic phase evolution produced via sintering process with increasing temperatures. The first stage was dominated by paramagnetic states with some superparamagnetic behavior; the second stage was influenced by moderately ferromagnetic states and some paramagnetic states; and the third stage consisted of strongly ferromagnetic states with negligible paramagnetic states. They found that three factors sensitively influenced the sample's content of ordered magnetism—the ferrite-phase crystallinity degree, the number of grains above the critical grain size and the number of large enough grains for domain wall accommodation.

In this research work, we investigated parallel evolution of microstructure-property relationship during the grain growth of nanosized starting powder. The question that we asked was how does the magnetic and electrical property evolve within the nanosized regime? Hence, cobalt zinc ferrite ($Co_{0.5}Zn_{0.5}Fe_2O_4$) was prepared by the mechanical alloying method and consequent changes in its microstructures, magnetic, electrical properties and their relationships were reported.

2. Experiment

2.1 Sample preparation

The starting powders of CoO (99%), Fe_2O_3 (99.5%) and ZnO (99.0%) were mixed according to the targeted proportion and were milled for 12 hours using a SPEX8000D mechanical machine employing a ball mill equipped with hardened steel vials and balls. Then, resulting powders were granulated with 1% PVA and lubricated using zinc stearate.

The granulated powder was pressed at 400MPa to obtain 3.0 gram toroidal sample shape. 16 toroidal samples were prepared in order to closely monitor the evolution of microstructure-property relationship of this work. The toroidal samples were sintered in air, with one sample sintered only once at one temperature, at different sintering temperatures from 600°C to 1350°C with 50° increment with 10 hours holding time. This route is called a multi-sample sintering route.

2.2 Characterization

The milled starting powder particles size and microstructure of the samples were measured by using a Scanning Transmission Electron Microscope (STEM) and a Field Emission Electron Microscope (FESEM). The measurements were carried out using the same machine (FEI NOVA NanoSEM 230). The microstructure was observed through machine and the grain size was measured by the mean linear intercept method of over 200 grains. The phase identification for the prepared powder and sintered samples were examined with X-ray diffraction (Philips Expert Pro PW3040) using CuKα.

Magnetic properties of the samples were investigated using MATS-2010SD Static Hysteresisgraph. The density of the sintered samples was measured using the Archimedes' method using distilled water as an immersion fluid.

The samples were coated with conducting silver paste on both sides for good contact between the surfaces for resistivity measurement. The samples were put between parallel plates and connected to a Keithley model 6485 low-current picoammeter, and the parallel plates were placed at the center of an electrical furnace. The measurement was carried out by measuring the samples current for every 5°C temperature interval from 100°C to 300°C. The resistance value could be calculated from the current value obtained by using the equation below:

$$V=IR \tag{1}$$

Where "R" is the electrical resistance, "I" is current, and "V" is the voltage. The value of voltage was fixed at 9 V. Then, the resistivity was determined by the equation below:

$$\rho= AR/l \tag{2}$$

Where "R" is the electrical resistance," A" is the cross sectional area and "l" is the height of the toroid sample, respectively.

3. Results and discussions

3.1 Particle size analysis

Fig. 1 (a) and (b) show the microstructure of the milled powder examined by STEM after 12 hours of milling and the grain-size distribution of the milled sample respectively. The size range of the particles varied inhomogenously from 0.013μm to 0.283 μm with an average particle size of 0.075 µm. The non-uniformity of the force of the milling media (steel balls and vials) on the powders during the high-energy milling process might be the cause of inhomogeneous particle sizes. However, this average particle was considered to be small for a starting powder, compared to that normally used (0.8 µm to 2 µm) in ferrite-production industry and even in much of the past research on ferrite. A narrow particle size range of the sample is important in order to obtain uniform grain size. Also, the bigger surface area of the powder, the greater driving force for densification so that higher densities could be achieved. According to Goldman (1990) [8], the optimization control of the starting particle size is important in order to control the evolution of the microstructure of the sample such as grain size, pores size and density because the magnetic properties are sensitive to these parameters.

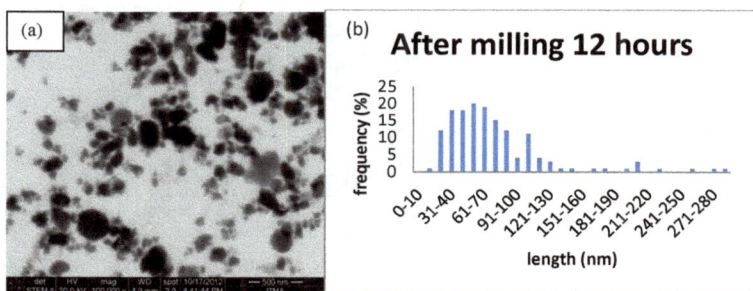

Figure. 1: (a)   STEM image of the as-milled $Co_{0.5}Zn_{0.5}Fe_2O_4$, (b) grain distribution of the as-milled $Co_{0.5}Zn_{0.5}Fe_2O_4$.

3.2   Phase analysis

Fig. 2 shows XRD spectra of the prepared and sintered sample from 600°C to 1350°C prepared via high energy ball milling. The signature peaks of the three starting materials, Fe_2O_3 at 2θ= 33.19°, ZnO at 2θ= 35.95°, CoO at 2θ= 36.95°, were evident for the as-milled sample, and can be indexed to ICDD cards of 01-073-2234 for Fe_2O_3, 01-089-2803 for CoO, and 01-089-0511 for ZnO. $Co_{0.5}Zn_{0.5}Fe_2O_4$ peaks appeared with the sintering as low as 600°C indicating the presence of crystalline cobalt zinc ferrite.

A complete phase of cobalt zinc ferrite was observed to form at 800°C sintering temperature with all diffraction peaks corresponding with the ICDD no. 89-1012 ($ZnFe_2O_4$) and 22-1086 ($CoFe_2O_4$). As the sintering temperature increased, the average grain size also increased, yielding more crystallization of the sample. At the same time, the magnetic phases were increasing due to the increasing crystalline mass having a cubic spinel structure. The increase of the magnetic phase in the samples caused the A–B interaction contribution in the spinel to be increased. Therefore, higher sintering temperature leads to higher values of magnetization.

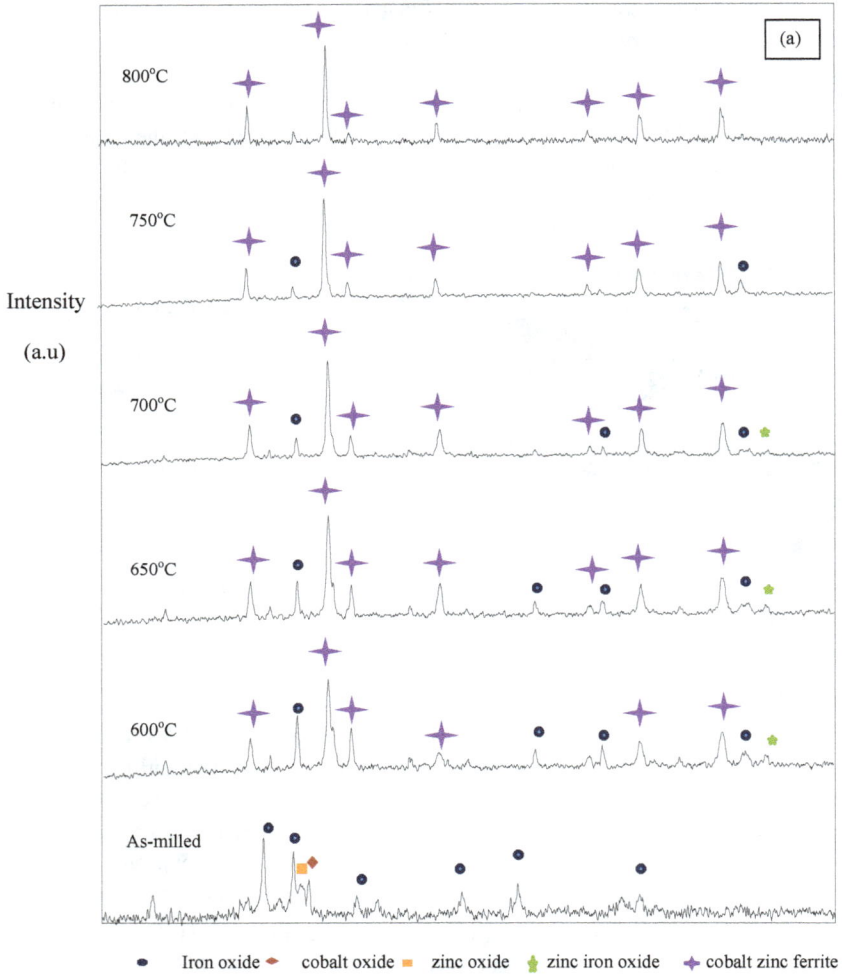

Figure 2. XRD pattern (a) for cobalt zinc ferrite sintered at 500°C to 800°C, (b) sintered at 850°C to 1350°C.

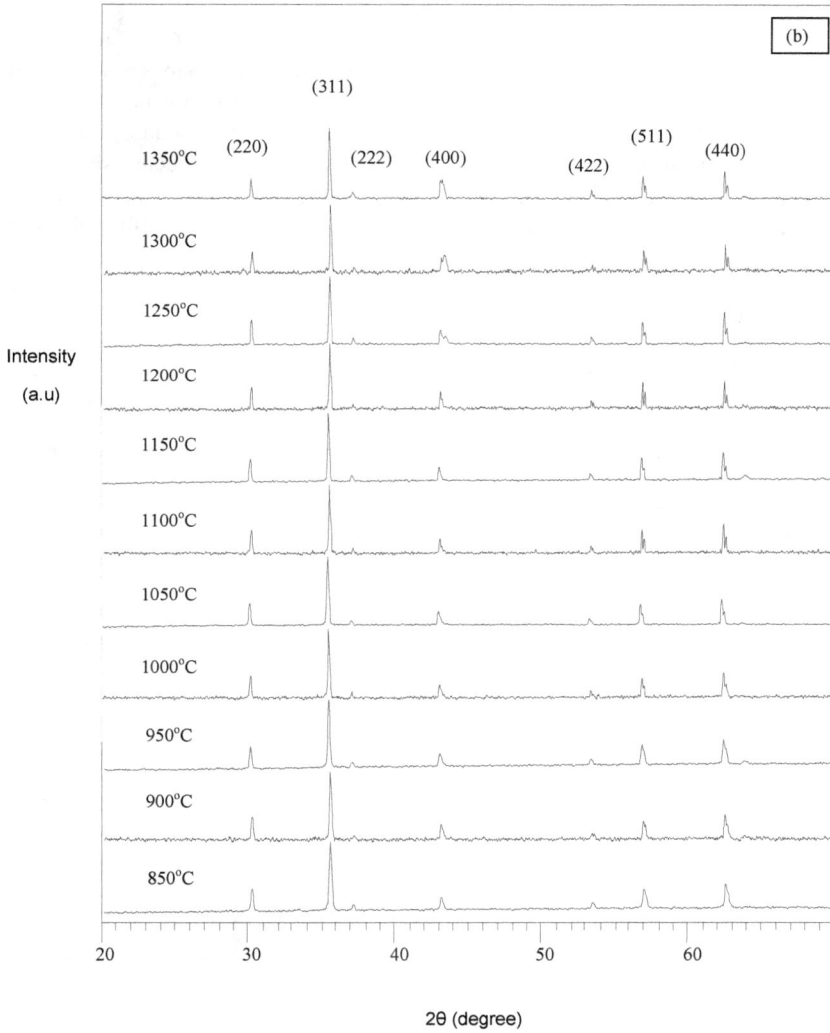

(Continue) Figure 2.   XRD pattern (a) for cobalt zinc ferrite sintered at 500°C to 800°C, (b) sintered at 850°C to 1350°C.

Table 1 shows the density and porosity of the samples sintered from 600°C to 1350°C. Samples sintered from 600°C to 750°C, showed low values of density caused by the incomplete formation of grains, which hindered the movement of domain walls. Meanwhile samples sintered from 800°C to 1250°C, showed an increase in density value due to increase of sintering temperature but slightly dropped after being sintered at 1300°C. The sample sintered at 1300°C showed the presence of pores which caused a decrease in density. The pores formation was subjected to the probability of the zinc loss and rapid grain growth at high sintering temperature, causing it to be trapped inside the grains. Table 1 shows where the porosity generally decreased up to the 1250°C sintering temperature and increased at the 1300°C sintering temperature.

Table 1. *Theoretical Density, Measured Density, Porosity and Average Grain Size for cobalt zinc ferrite.*

Temperature (°C)	Density Measured (g/cm³) (±0.01)	Theoretical Density (%) (±0.01)	Porosity (%) (±0.01)	Average Grain Size (µm) (±0.001)
600	4.73	89.27	10.73	0.095
650	4.79	90.48	9.51	0.098
700	4.77	89.91	10.09	0.106
750	4.75	89.35	10.64	0.107
800	4.41	83.08	16.92	0.121
850	4.46	87.23	12.77	0.126
900	4.64	87.27	12.73	0.183
950	4.69	92.85	7.15	0.216
1000	4.70	88.36	11.64	0.258
1050	4.57	85.94	14.06	0.272
1100	4.73	88.92	11.08	0.319
1150	4.83	90.63	9.37	0.508
1200	4.81	90.48	9.52	0.925
1250	4.83	90.87	9.13	2.402
1300	4.64	87.01	12.99	2.764
1350	4.00	74.79	25.24	3.340

3.3 Magnetic properties

Fig. 3 shows magnetic induction, B, against magnetic field strength, H, (B-H hysteresis loop) plotted for different sintering temperatures. According to the previous report, the shape of the hysteresis-loop depended on the grains in the sample [8] and our results showed a good agreement with the reference [8]. There were three groups observed based on different magnetic hysteresis loop shapes in Fig. 3. The first-group was samples sintered at 600°C to 800°C, a small hysteresis was observed indicating the ferromagnetic nature of the material. The linear-looking loops have very low saturation induction, B_s indicating a very small amount of a ferromagnetic phase. However, a significant coercivity, H_c, with somewhat elongated shape was due to necking. Even though, the XRD pattern for sintered sample at 800°C showed only the presence of cobalt zinc ferrite peaks, the Bs value was still very low, indicating that the crystalline phase was not yet dominant.

The samples sintered at 850°C to 1050°C exhibited a mixture of single-domain and multi-domain grains and it was classified as a second-group. Higher B_s values but falling H_c values (Fig.4) showing that higher ferromagnetic phase crystallinity and starting of the dominance of multi-domain magnetization-demagnetization process. The third-group belongs to the samples sintered from 1100°C to 1350°C having a well-defined sigmoid shape because of very high crystallinity, large grain size and high density which allowed domain walls to move with great ease in the magnetization and demagnetization process. Table 2 shows the variation trend in the coercivity, H_c values for the sintered samples. The increasing variation of H_c values with increasing grain size, D was observed for sample sintered at 600°C and 850°C. Meanwhile for samples sintered at 900°C and above, H_c values were decreasing with increasing grain size, D. From previous study by Sanchez et al. (2002) [9], the H_c was proportional with 1/D, therefore the H_c decreased with increasing sintering temperature. The reason for this was that in small particles the formation of a closed magnetic flux becomes energetically less favourable so that the magnetic domain size with uniform magnetization became more and more identical with the grain size. This grain size is defined as the critical threshold where the multidomain materials change to a monodomain material. From Fig. 4, a similar trend was observed for the sintered samples where the grain size was reduced, it was found that before the H_c was increased to a maximum value and then decreased towards zero value. The sintered samples showed a maximum H_c value when sintered at 850°C which indicated the critical grain size, corresponded to the transition of domain from single-domain to multi-domain. Based on our study, the critical size for the sintered sample was 0.126 µm.

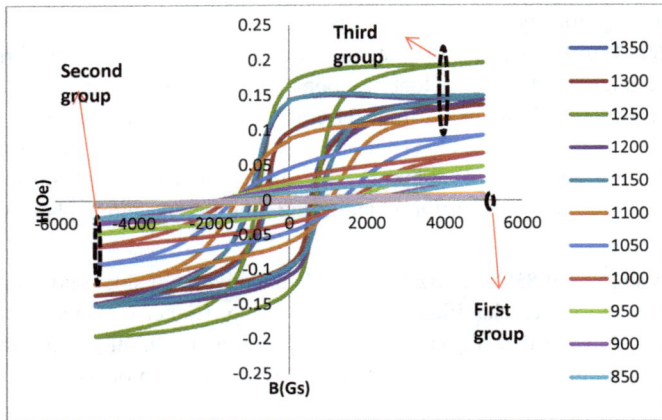

Figure 3.   *B-H Hysteresis loops for the multi-sample sintering from 600°C to 1350°C.*

The samples with grain size larger than the critical size are multi domain, meanwhile samples with grain size with smaller than critical sizes are single domain.

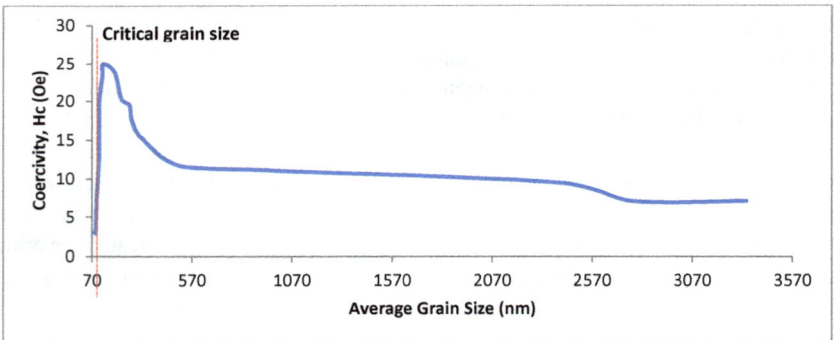

Figure 4.   *Coercivity versus average grain size for sintered sample from 600°C to 1350°C*

Table 2.   *Induction, B_s, and Coercivity, H_c, of the sintered sample*

Sintering Temperature (°C)	Coercivity, Hc (Oe) (±0.01)	Induction, B_s(Gauss) (at 5000 Oe)
600	7.58	35.75
650	8.81	37.26
700	13.12	55.35
750	16.56	65.55
800	23.10	138.20
850	24.93	259.50
900	23.84	336.50
950	20.26	482.00
1000	19.48	671.80
1050	17.40	929.60
1100	15.26	1215.00
1150	11.70	1330.00
1200	11.12	1503.00
1250	9.47	1965.00
1300	7.03	1367.00
1350	7.02	1486.00

*Critical grain size

3.4   Resistivity dependence

The temperature dependence of electrical resistivity was measured in the temperature range of 373 K to 573 K as shown in Fig. 5. From the results, the resistivity could be classified in two parts based on their shape of the graph. The first-part (Fig. 5) was categorized for the samples sintered from 600°C to 750°C. It showed the resistivity values increasing with the rise of the temperature from 373 K and attained the maximum value at a particular temperature and decreased with a further increase in temperature. Two considerations were made to explain this phenomenon, which were due to the small size of the starting powder and the incomplete phase of the ferrites. The size of the particles ranged from 0.085 μm to 0.107 μm (Table 1); it is not easy for electrons/holes to move when giving external energy, thus presenting high resistivity inside the sample. The second consideration of this phenomenon was due to the incomplete phase of the ferrite.

The second part of resistivity was the sample sintered from 800°C and above. Samples which belong to this second group showed a decreasing value of resistivity with increasing temperature. The decrease in resistivity value was speculated to be due to the excess of electrons released from sites with Co atoms reduced to Co^{2+} i.e. to their lowest valence and also from sites with Fe^{2+} ions. According to Rezlescue model [10,11] the exchanging of electrons between Fe^{2+} and Fe^{3+} ions and that of holes between Co^{3+} to Co^{2+} ions may be the likely conduction mechanism:

$$Co^{3+} \longleftrightarrow Co^{2+} \text{ (Hole Conduction) and} \qquad (3.1)$$

$$Fe^{2+} \longleftrightarrow Fe^{3+} \text{ (Electrons Conduction)} \qquad (3.2)$$

As sintering temperature were increased, bigger grain size increases the grain-to-grain contact area and caused the electron to flow, therefore lowering the resistivity of the sample. A study by Viswanathan, and Murthy in 1990 has been carried out in which two regions of conductivity were observed [12]. One region was of low conductivity containing Co^{2+} and Co^{3+} ions and the other region was of high conductivity containing Fe^{2+} and Fe^{3+} ions.

The graph obtained (Fig. 5) showed the resistivity values decreased with increasing temperature and this confirmed that the sintered samples under investigation possessed semiconducting behavior.

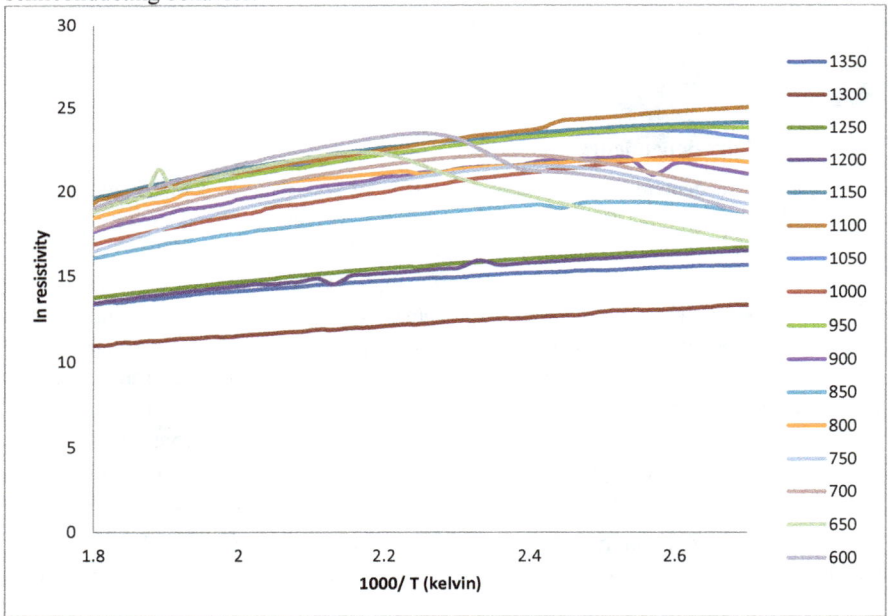

Figure 5. *The temperature dependence of electrical resistivity for sintered sample from 600°C to 1350°C*

3.5   Microstructure study

The surface microstructure of toroidal bulk $Co_{0.5}Zn_{0.5}Fe_2O_4$ compacts with sintering temperature from 600°C to 1350°C as shown in Fig. 6. After the sintering process, samples were directly viewed under FESEM without polishing to study the microstructures. The grain size distribution and average grain size are shown in Fig. 7 and Table 1. It shows that the average grain size were increased from 0.095µm to 3.340µm, this increase indicated the microstructural evolution due to grain growth of the samples.

Sample with low sintering temperature (600°C) (Fig. 6) showed the effect of mechanical alloying after the sintering process and also, produced average grain size distribution of 0.095µm. The microstructure of sample sintered at 650°C, showed clearly smooth surface of the grain showing the effect of sintering parameter. We can also observe a necking process occurring within the sample, and attributed it to the diffusion of materials. Comparing Fig. 6 to the X-ray Diffraction result in Fig. 2, it showed that $Co_{0.5}Zn_{0.5}Fe_2O_4$ phase just started to appear. Sample sintered at 800°C showed that the average grain size was larger compared to the samples sintered at 700°C and 750°C. Samples with sintering temperature of 800°C to 950°C produced average grain size distribution in the range of 0.12µm to 0.22µm. The samples sintered from 800°C to 950°C mostly showed the process of necking, indicating the growth of grains over the sintering. The samples sintered at 1000°C to 1250°C (Fig. 6) showed increasing grain size, from 0.25 µm to 2.40 µm. At this stage, grain boundaries (and grains) were formed. Samples sintered at 1300°C and 1350°C (Fig. 6) however showed the existence of intragranular pores in the grains due to rapid grain growth. Intragranular pores are known to be bad inclusions because the presence of these pores pin down the magnetic moment in grains thus reduces the magnetization.

Fig. 7 showed histograms of grain size distribution of the sintered samples. The grain size shifted to the larger grain size as the sintering temperature increased. Based on the coercivity results (Fig. 4), the transition from 850°C to 900°C sintering temperature showed a drastic fall of the coercivity value.

Therefore, a grain size range of 0.121µm to 0.130µm was chosen as the range of critical size for single domain particles. Based on histograms, the left side from red line is grain size with less than 0.130 µm. With increasing sintering temperature, the area was reduced and completely disappeared for samples sintered at 1200°C and up to 1350°C. This shows that the number of grain size exceeding the single domain to multi domain critical grain size were increased, Hence the number of domain walls were also increased, therefore the contribution of domain movement to ease the magnetization were increased.

Figure 6. *Microstructure of cobalt zinc ferrite sintered at (a) 600°C, (b) 650°C, (c) 700°C, (d) 750°C, (e) 800°C, (f) 850°C.*

Figure 6.   Microstructure of cobalt zinc ferrite sintered at (g) 900°C, (h) 950°C, (i) 1000°C, (j) 1050°C, (k) 1100°C, (l) 1150°C, (m) 1200°C, (n) 1250°C.

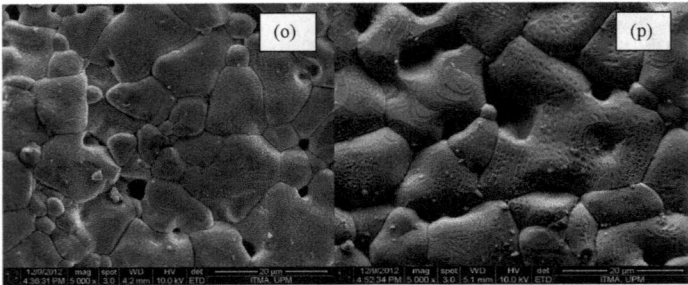

Figure 6.   *Microstructure of cobalt zinc ferrite sintered at (o) 1300°C, (p) 1350°C.*

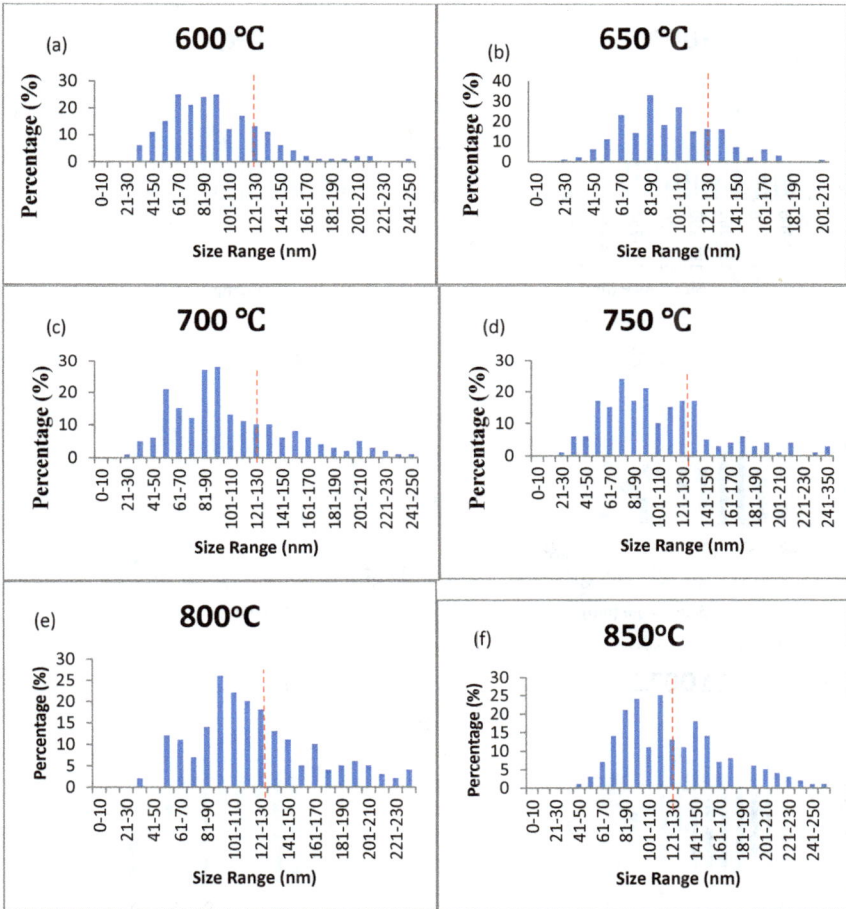

Figure 7.   Grain-size distribution of cobalt zinc ferrite sintered at (a)600°C, (b) 650°C, (c) 700°C, (d) 750°C, (e) 800°C, (f) 850°C.

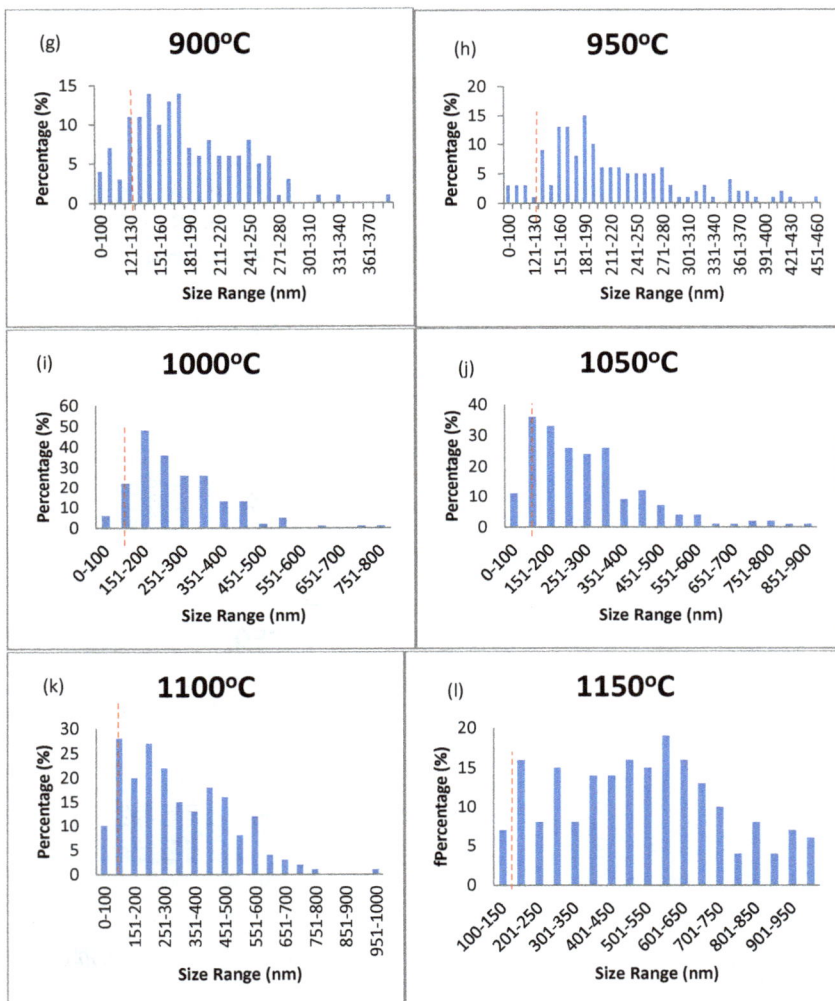

(Continue) Figure 7. Grain-size distribution of cobalt zinc ferrite sintered at (g) *900°C, (h) 950°C, (i) 1000°C, (j) 1050°C, (k) 1100°C, (l) 1150°C.*

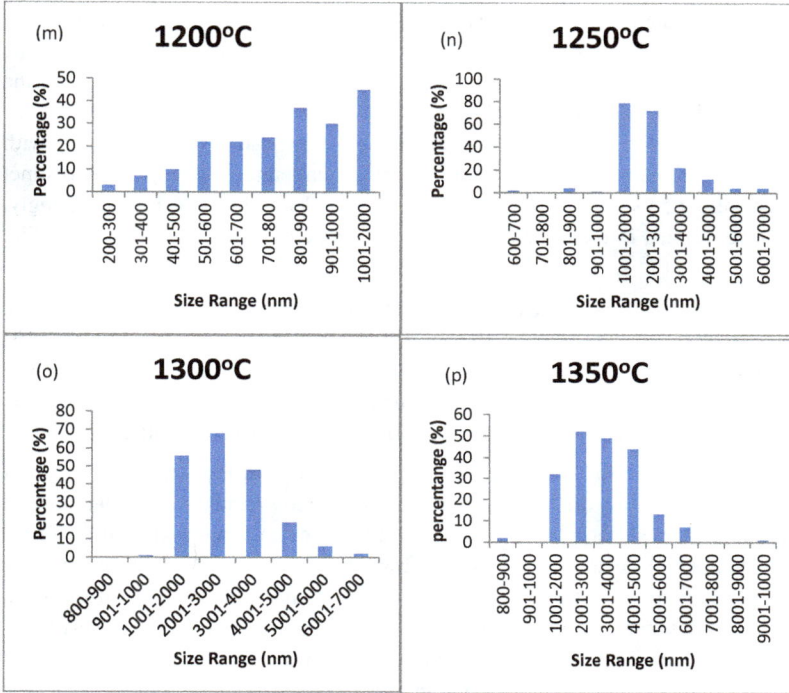

*(Continue)Figure 7. Grain-size distribution of cobalt zinc ferrite sintered at (m) 1200°C,
(n) 1250°C, (0) 1300°C, (p) 1350°C.*

3.6   Activation energy

By using equation 3.3 from the behavior of particle grain growth, the activation energy of
grain growth can be predicted as mentioned in coble's theory [13]

$$d\ln k /(dT) = RT^2 \qquad (3.3)$$

Where k is the specific reaction rate constant, activation energy is referred as Q, T is the
absolute temperature, and R is the ideal gas constant. The value of k can be directly be
related to grain size, which results in equation 3.4 below [14]:

$$\text{Log } D = (-Q/2.303R)1/T + A \qquad (3.4)$$

Where T is the absolute temperature, A is the intercept and D is the grain size. By using equation 3.4, three best-fitted straight-line plots were obtained for our samples where the plots of log D versus the $1/T$ are shown in Fig.8. From equation 3.4, the slope of the line as $-Q/2.303$ R and the value of the activation energy of grain growth can be calculated from the Arrhenius plot. The activation of energy shows an increasing value with increasing grain size caused by increasing sintering temperature, giving three distinct values which were 7.58 kJ/mol, 47.11 kJ/mol and 189.54 kJ/mol (Table 1). Interestingly, the slopes corresponded to the three grouping of magnetic properties shown earlier (Fig. 3).

One consideration was made to explain this phenomenon, which corresponds to the mechanisms of sintering. There are different sintering mechanisms, or in other words, different modes of matter transport from sources (surfaces, grain boundaries, defect) to the sinks (bridge connection) [15]. The different mechanisms are surface diffusion, volume diffusion, vapor transport and grain boundary diffusion (intergranular diffusion).

In the case of cobalt zinc ferrite sample, portions 1, 2 and 3 in Fig.8 are referred to as diffusion of processes with different activation energy requirements. Speculatively, to transport material from surfaces, all mechanisms are active and it referred to diffusion on the surface (number 1).To transport material from the interior of the grain boundaries both volume diffusions and grain boundaries are active and it referred to diffusion on grain boundaries (number 2).To transport material from intra-granular defects, this is the only active mechanism and it referred to volume diffusion (number 3). As the crystalline structure is disturbed at the interfaces of the grains, the activation energies for diffusion are smaller for solid interfaces than for scattering into volume. Generally, the temperature increases the rate of all sintering mechanism. Generally $Q_{surface} < Q_{grain\ boundaries} < Q_{volume}$ as higher temperature accelerates volume diffusion compared to interfacial diffusion [15].

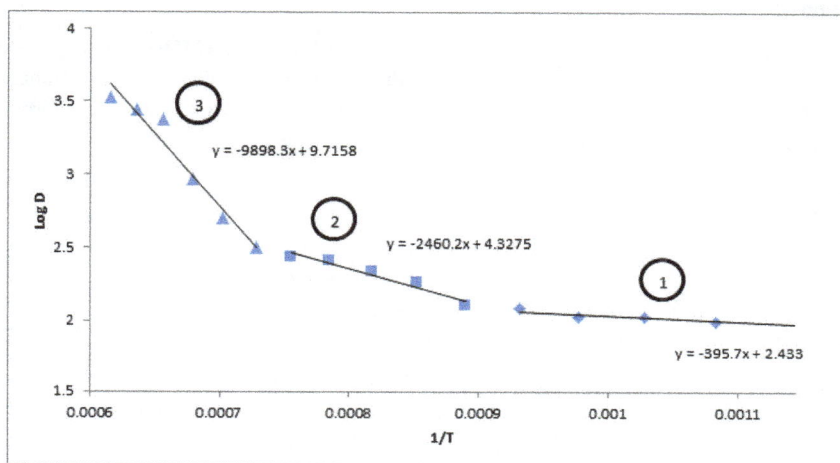

Figure.8.   *Plots of log D versus the reciprocal of absolute temperature (1/T).*

Table 3.   *The Activation Energy for sintered sample of cobalt zinc ferrite.*

Group (Temperature)	Activation energy, kJ/mol
1 (600 °C -800 °C)	7.58
2(850 °C-1050 °C)	47.11
3(1100 °C – 1350 °C)	189.54

Conclusions

Cobalt zinc ferrite ($Co_{0.5}Zn_{0.5}Fe_2O_4$) was synthesized using the mechanical alloying method and the effect of sintering temperature on its microstructure, magnetic and electrical properties were studied. We found that $Co_{0.5}Zn_{0.5}Fe_2O_4$ cannot be formed directly through milling alone, but heat treatment was necessary for its formation. It was possible to begin the formation of $Co_{0.5}Zn_{0.5}Fe_2O_4$ by sintering the sample at 600°C giving grain sizes in the nanometer range.

The shape of the magnetic hysteresis loops of cobalt zinc ferrite is now clearly understood: a manifestation of the combined effect of a particular set of conditions such as phase purity level, degree of crystallinity and microstructural factors, especially average grain size and density/porosity give a particular pattern or shape of magnetic hysteresis. The samples showed decreasing value of resistivity with increasing temperature and we have confirmed that, the samples under investigation possessed semiconducting behavior i.e. falling resistivity with rising temperature.

Acknowledgement

The researchers wish to thank the Ministry of Higher Education (MOHE) Malaysia for providing an LRGS research grant, Universiti Putra Malaysia, Malaysia for the graduate Research Fellowship and NanoMag (Nanostructured Polycrystals and Magnetism Group) for support of this project.

Reference

[1] Qu, Y., Yang, H., Yang, N., Fan, Y., Zhu, H., Zou, G.,The effect of reaction temperature on the particle size, structure and magnetic properties of coprecipitated CoFe2O4 nanoparticles, Mater. Lett., 60, 3548-3552(2006). http://dx.doi.org/10.1016/j.matlet.2006.03.055

[2] Sun, K., Z. Ian, S. Chen, Y. Sun and Z. Yu,Effect of sintering process on microstructure and magnetic properties of high frequency power ferrite, Rare Metals,25(2006)., 509-514. http://dx.doi.org/10.1016/S1001-0521(07)60135-1

[3] Azadmanjiri, J. Structural and electromagnetic properties of Ni–Zn ferrites prepared by sol–gel combustion method J. Materials Chemistry and Physics, 109(1) (2008)., 109-112. http://dx.doi.org/10.1016/j.matchemphys.2007.11.001

[4] Gul I. H., Abbasi a. Z., Amin F., Anis-ur-Rehma M., and Maqsood a. Structural, magnetic and electrical properties of Co1−xZnxFe2O4 synthesized by co-precipitation method Journal of Magnetism and Magnetic Materials, 311(2) (2007), 494-499. http://dx.doi.org/10.1016/j.jmmm.2006.08.005

[5] Sharifi I and Shokrollahi H. Nano structural, magnetic and Mossbauer studies of nano sized Co1_xZnxFe2O4 synthesized by co-precipitation. Journal of Magnetism and Magnetic Materials, 324(15) (2012)., 2397-2403. http://dx.doi.org/10.1016/j.jmmm.2012.03.008

[6] Idza I. R., Hashim M., Rodziah N., Ismayadi I., and Norailiana a. R. Influence of evolving microstructure on magnetic-hysteresis characteristics in polycrystalline nickel–zinc ferrite, Ni0.3Zn0.7Fe2O4.Materials Research Bulletin, 47(6) (2012), 1345-1352. http://dx.doi.org/10.1016/j.materresbull.2012.03.007

[7] Ismail, I. M. Hashim, I.R. Ibrahim, R. Nazlan, F. MohdIdris, S.E. Shafie, M. Manap, G. Bahmanrokh, N.H. Abdullah, W.N. Wan Rahman, Crystallinity and magnetic properties dependence on sintering temperature and soaking time of mechanically alloyed nanometer-grain Ni0.5Zn0.5Fe2O4, Journal of Magnetism and Magnetic Materials, Vol (333), 2013, 100-107

[8]   C.P. Bean, Hysteresis loops of mixtures ferromagnetic micropowders. J. Appl. Phys. 26 (1955) 1381–1383. http://dx.doi.org/10.1063/1.1721912

[9]   Sánchez, R.D., Rivas, J., Vaqueiro, P., López-Quintela, M.A., Caeiro, D., Particle size effects on magnetic properties of yttrium iron garnets prepared by a sol–gel method, Journal of Magnetism and Magnetic Materials, 247 (1) (2002), 92–98. http://dx.doi.org/10.1016/S0304-8853(02)00170-1

[10]   Mott, N. F.. 1. The model for hopping conduction in glasses (1968).

[11]   Khan, H. M.. Materials Sciences and Applications, 02(08)(2011), 1083-1089. http://dx.doi.org/10.4236/msa.2011.28146

[12]   Viswanathan, B. and Murthy, V.R.K., Ferrite Materials: Science and Technology, Narosa Publishing House(1990).,

[13]   Coble, R.L.,Sintering crystalline solids, 1. Intermediate and final state diffusions models.J. Appl. Phys. 32(1961), 787-792. http://dx.doi.org/10.1063/1.1736107

[14]   Jarcho, M., Bolen, C.H., Thomas, M.B., Bobick, J., Kay, J.K., Doremus, R.H.,Hydroxylapatite synthesis and characterization in dense polycrastalline form.J. Mater Sci11 (1976)., 2027-2035. http://dx.doi.org/10.1007/PL00020328

[15]   Sandra Galmarini. Work practices "Ceramics Process": TP3 Sintering 1-15(2011).

CHAPTER 8

Effect of the sintering temperature on the microstructure and optical properties of ZnO ceramics

B. Subha[1], R. Saravanan[3*] and N. Srinivasan[2]

[1]PG Department of Physics, E.M.G Yadava Women's College, Madurai – 625014, Tamil Nadu, India

[2]Research Centre and PG Department of Physics, Thiagarajar College, Madurai – 625009, Tamil Nadu, India

[3]Research Centre and PG Department of Physics, The Madura College, Madurai – 625011, Tamil Nadu, India

Email:subi_shini_phy@yahoo.com; vasan692000@yahoo.co.in; *saragow@gmail.com

Abstract

ZnO (alfa aesar, 99.9%) powder was sintered in a conventional furnace at different temperatures (200°C, 800°C and 1000°C). The structural study was done by powder X-ray diffraction and it was found that there were no other phases present in the material. The crystallite size of the sintered ZnO powder was found to increase with increasing sintering temperature. The effect of sintering on grain growth was investigated by scanning electron microscopy (SEM). SEM reveals that the average grain size increases with increasing sintering temperature. The optical band-gap E_g of sintered ZnO has been determined from UV-Vis absorption spectra of these samples and the band gap was found to decrease with increase in sintering temperatures. In the present work, the effect of sintering temperature and hold time on the microstructure and optical properties of ZnO ceramics has been carried out.

Keywords

Zinc oxide, Sintering, Optical Energy Gap, Rietveld Analysis, Maximum Entropy Method.

Contents

1. Introduction

Nanocrystalline materials have attracted wide attention due to their unique properties and immense applications in nano device fabrication. Zinc oxide (ZnO) is vastly used for the development of optoelectronic devices like light emitting devices and solar cells. Nano zinc oxide is non-toxic, with wide band gap and has been identified as a promising semiconductor material for exhibiting ferromagnetism (RTFM) at room temperature. All these properties make ZnO, a potential material in the field of nanotechnology [1].

The conventional sintering process has the advantages of being cheap, easy-to-use, safe and able to be implemented in any scientific laboratory. The present article is devoted to examine the influence of sintering temperature on the microstructure and optical properties of ZnO.

2. Experimental

ZnO (Alfa aesar, 99.9%) powder was sintered in a conventional furnace at different temperatures (200°C, 800°C and 1000°C) for 3 hrs. These sintered samples were characterized by XRD, SEM, EDAX and UV – Vis techniques.

3. Results and Discussion

3.1 XRD Characterization

Fig.1 shows the XRD patterns of sintered ZnO samples. The XRD patterns reveal that there are no other phases present in the sintered samples.

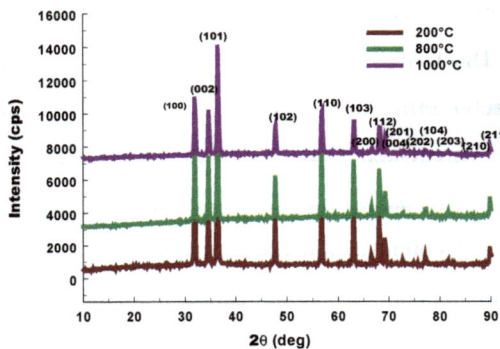

Figure 1.   XRD patterns of ZnO sintered at different temperatures.

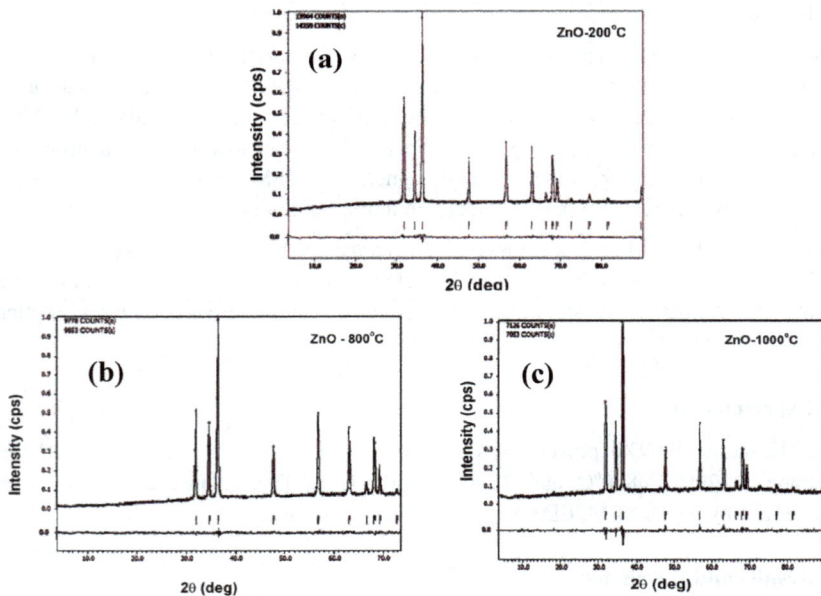

Figure 2.   Refined powder profile of Sintered ZnO at (a) 200°C, (b) 800°C and (c) 1000°C.

The ZnO nanostructure was analyzed further using the Rietveld refinement method [2]. JANA 2006 software [3] was used to fit the experimental diffraction patterns by considering the hexagonal structure of w-ZnO, which belongs to the $P6_3mc$ space group containing two formula units per primitive cell. The input atomic coordinates (x, y, z) were set as $\left(\frac{1}{3}, \frac{2}{3}, 0\right)$ and $\left(\frac{2}{3}, \frac{1}{3}, \frac{1}{2}\right)$ for Zn atoms, and $\left(\frac{1}{3}, \frac{2}{3}, u\right)$ and $\left(\frac{2}{3}, \frac{1}{3}, u+\frac{1}{2}\right)$ for O atoms, where $u = 0.3875$ [4]. The experimental and calculated XRD patterns of ZnO sintered at different temperatures are shown in fig. 2 (a, b, c). The refined parameters are summarized in Table 1.

Table 1.   *Refined structural parameters of ZnO at different temperatures.*

Refined Parameter	Sintering Temperature		
	200°C	800 °C	1000°C
R_p (%)	3.43	3.35	3.80
wR_p (%)	4.65	4.49	4.77
R_{obs} (%)	1.12	1.11	1.87
wR_{obs} (%)	1.02	1.21	2.01

The grain size of the sintered ZnO samples was estimated from the observed full width at half maximum (FWHM) of the XRD peaks using GRAIN software [5]. The variation of FWHM, d-spacing and grain size with sintering temperature are shown in fig. 3, 4 and 5 respectively. The FWHM of ZnO slightly decreases with sintering temperature which indicates a structural ordering of ZnO lattice inside the grain and grain boundaries, and also a structural disordering of impurities also at the grain boundaries in the ZnO ceramic [6]. The grain size increases with increase in sintering temperature while the d-spacing decreases.

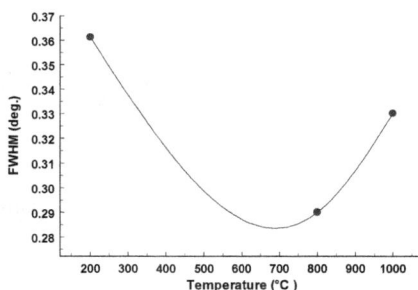

Figure 3.   *The variation of FWHM with sintering temperature.*

Table 2.   *The grain size of the ZnO sintered at different temperatures.*

Temperature(°C)	d-Spacing (Å)	Grain Size (10^{-10}m)
200°C	2.46873	230.8754
800°C	2.46677	284.2514
1000°C	2.46437	252.9058

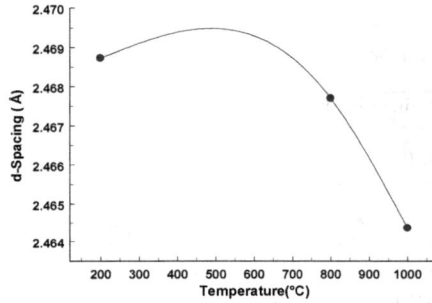

Figure 4.   *The variation of d-spacing with sintering temperature.*

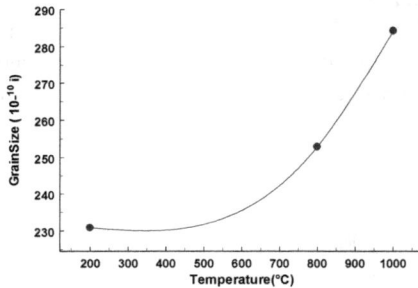

Figure 5.   *The variation of grain size with sintering temperature.*

Maximum entropy method (MEM) [7] is an exact and versatile tool to determine the electron density distribution of crystalline structures. This method needs only minimum information to determine the spatial electron density distribution in a solid crystal with high accuracy based on probabilistic approaches, yielding least biased information. In this work, the software package PRIMA [8] was used for MEM computations. MEM Refined parameter of ZnO sintered at different temperatures are tabulated in Table 3.

Table 3.   *MEM Refined parameter of ZnO sintered at different temperatures.*

Refined Parameter	Sintering Temperatures		
	200°C	800 °C	1000°C
No. of Cycles	3933	3498	3482
$R_{(MEM)}$ (%)	0.4285	0.42388	0.4447
$wR_{(MEM)}$ (%)	0.4704	0.4687	0.4773

3.2   SEM Measurements

The SEM micrographs of ZnO sintered at different temperatures are shown in Fig. 6. Table 2 shows the relationship between the sintering temperature and lattice spacing. When the sintering temperature increases, there is a rapid decrease in the density of vacant lattice sites and local lattice disorders and the volume of the unit cell goes towards expected normal values. Hence, the particle size increases with sintering temperature (Fig. 7). The particle sizes of ZnO sintered at different temperatures are listed in Table 4.

(a)ZnO-200°C

(b)ZnO-800°C   (c)ZnO-1000°C

 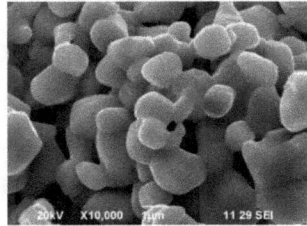

Figure 6.   *SEM images of ZnO sintered at (a) 200°C, (b) 800°C and (c) 1000°C.*

Table 4.   *Particle size of ZnO sintered at different temperatures.*

Temperature(°C)	Particle Size (nm)
200°C	344
800°C	900
1000°C	1462

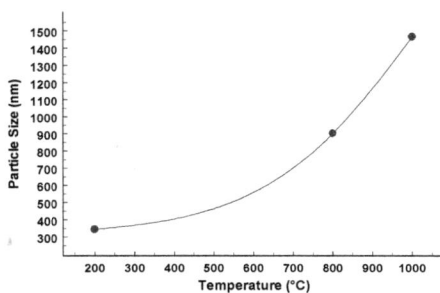

Figure 7.   *Size of the sintered ZnO at different temperatures.*

3.3   EDAX Analysis

The elemental composition of the sintered ZnO samples are determined by EDAX and presented in Table 5. It is worth to note that EDAX estimation of O at% decreases with sintering temperatures and Zn at% increasing with sintering temperatures [6].

Table 5.   *EDAX analysis of the Zinc oxide samples at different temperatures.*

Sintering temperature(°C)	O (at %)	Zn (at %)
200°C	50.98	49.02
800 °C	45.28	54.72
1000°C	44.37	55.63

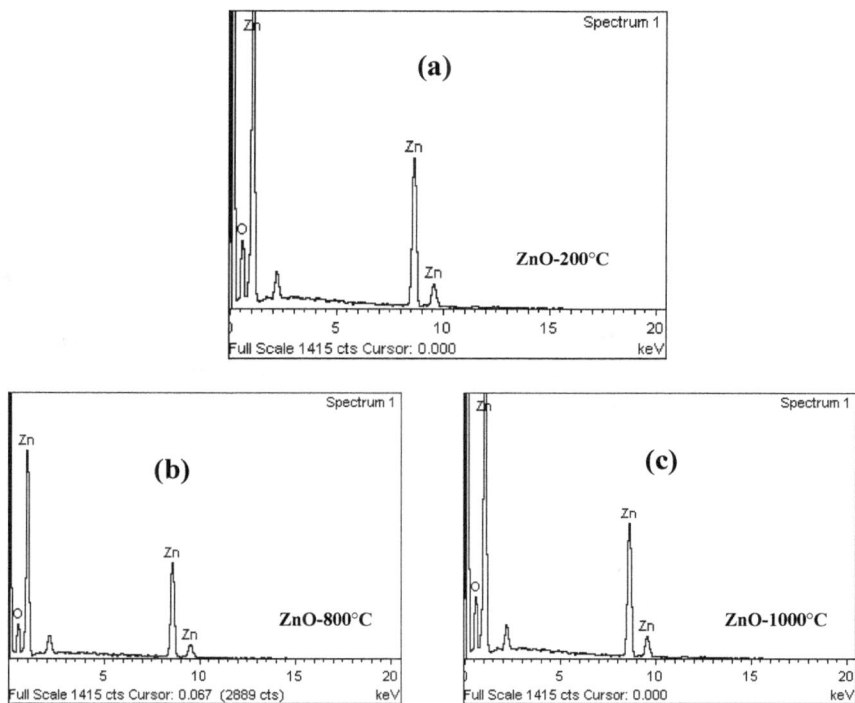

Figure 8. *EDAX analysis of sintered ZnO at (a) 200°C, (b) 800°C and (c) 1000°C.*

3.4 UV-VIS Measurement

The optical absorption spectra of sintered ZnO samples were recorded. The spectra show the absorbance peak in the range of 300 – 350nm. The optical band-gap E_g was determined by using Tauc's plot [9]. The Tauc's plot for sintered ZnO at different temperatures is shown in fig. 9. In ZnO ceramics, the optical band gap energy decreases with increase of sintering temperature. The band gap values of the sintered ZnO are 3.22 eV at 200°C, 3.19 eV at 800°C and 3.18 eV at 1000°C. Thus E_g decreases with increasing sintering temperatures as shown in fig. 10.

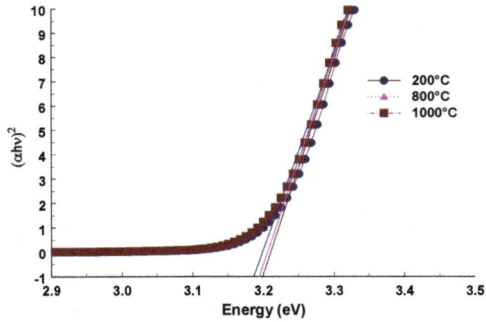

Figure 9.   Tauc's plot of ZnO sintered at different temperatures.

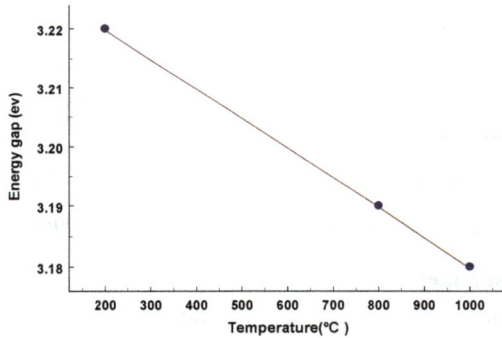

Figure 10.   The variation of optical band-gap with sintering temperature.

Conclusions

The effect of sintering temperature on ZnO ceramics was investigated in the temperature range of 200°C to 1000°C. The sintering temperature influences the microstructure and optical properties of the ceramics. The grain size increases with sintering temperature and the optical energy gap decreases with sintering temperature.

References

[1]   Ruby Chauhan, Ashavani Kumar, Ram Pal Chaudhary, Synthesis and characterization of silver doped ZnO nanoparticles, Arch. Appl. Sci. Res., 2 (5) (2010) 378- 385.

[2]   H. M. Rietveld, A Profile Refinement Method for Nuclear and Magnetic Structures, J. Appl. Crystallogr. 2 (1969) 65. http://dx.doi.org/10.1107/S0021889869006558

[3]   V. Petricek, M. Dusek, L. Palatinus, The crystallographic computing system, Institute of Physics, Praha, Czech Republic, 2006.

[4]   U. Ozgur, Y. I. Alivov, C. Liu, A. Teke, M. A. Reshchikov, S. Dogan, V. Avrutin, S. J. Cho, H. Morkoc, A comprehensive review of ZnO materials and devices, J. Appl. Phys. 98 (2005) 041301. http://dx.doi.org/10.1063/1.1992666

[5]   R. Saravanan, Doping-induced electron density modification at lattice sites of ZnO:Ga nanostructures: effects on vibrational and optical properties, J Mater Sci 49 (2014) 5529–5536. http://dx.doi.org/10.1007/s10853-014-8242-z

[6]   Mariem Chaari, Adel Matoussi, Zouheir Fakhfakh, Structural and Dielectric Properties of Sintering Zinc Oxide Bulk Ceramic, Laboratory of Composite Ceramic and Polymer Materials, Scientific Faculty of Sfax, Tunisia, Africa, (2011) 765-770.

[7]   L.B. McCusker, R. B. Von Dreele, D. E. Cox, D. Louer, P. Scardi, Rietveld refinement guidelines, J. Appl. Crystallogr. 32 (1999) 36. http://dx.doi.org/10.1107/S0021889898009856

[8]   F. Izumi, R.A. Dilanian, Research Developments in Physics, F. Izumi and R. A. Dilanian, "Recent Research Developments in Physics," Transworld Research Network, Trivandrum (ISBN 81-7895-046-4), 3 (2002) 699-726.

[9]   M. G. M. Sabri, B. Z. Azmi, Zahid Rizwan, M. K. Halimah, M. Hashim, M. Zaid, Effect of temperature treatment on the optical characterization of $ZnO-Bi2O_3-TiO_2$ varistor ceramics, International Journal of the Physical Sciences 6(6) (2011) 1388-1394.

CHAPTER 9

Structural characterization of beryllium and indium oxide powders

R. Yuvakkumar[1,2*], V. Milton[1], G. Ravi[1], S.I. Hong[2*]

[1]Nanomaterials Laboratory, Department of Physics, Alagappa University, Karaikudi - 630 004, Tamil Nadu, India

[2]Department of Nanomaterials Engineering, Chungnam National University, Daejeon, 305-764, South Korea

Email: yuvakkumar@gmail.com, sihong@cnu.ac.kr

Abstract

The effect of processing parameters such as reaction time, incubation and calcination temperature on the structural properties of beryllium and indium oxide powders employing a green synthesis technique has been investigated. A possible mechanism to understand the formation of metal-ellagate complex formation has been explored.

Keywords

Beryllium Oxide, Indium oxide, Green Synthesis, Structural Properties, Metal-Ellagate, Clacination

Contents

1.   Introduction

Beryllium oxide (BeO) is used in a variety of applications due to its unusual combination of optical, thermal, dielectric and mechanical properties [1]. Beryllium oxide material is applied in a wide range of applications such as high power devices or high density electronic circuits for high speed computers [2]. It may be used as windows, radomes and antennas for microwave communication systems and microwave ovens where both high thermal conductivity and electrical resistivity is needed [3]. Recently, Petrenko and his co-workers investigated the thermoluminescence and low-temperature luminescence properties of beryllium oxide [4]. Dawahra and his co-workers explored the investigation of BeO as a reflector for low power research reactor [5]. Fathalian and his co-workers revealed the beryllium oxide nanotube bundle as a gas sensor [6]. In addition, Fathalian and his co-workers investigated the optical properties of BeO nanotubes: Ab initio study [7]. Baima and his co-workers investigated beryllium oxide nanotubes and their connection to the flat monolayer [8]. Anota and his co-workers studied the electronic properties of functionalized (5,5) beryllium oxide nanotubes [9]. Wrasse and his co-workers studied the first principles study of native defects in BeO [10]. Wu and his co-workers studied the electronic and magnetic properties and structural stability of BeO sheet and nanoribbons [11]. Wang and his co-workers studied the growth of BeO nanograins synthesized by the polyacrylamide gel route [12]. Moreover, Wang and his co-workers studied the synthesis and sintering of beryllium oxide nanoparticles [13].

Indium oxides (In_2O_3) have attracted much attention in recent years because of their unique structural and optical properties [14-15]. Indium oxide, an n-type direct wide band gap transparent semiconductor have found a variety of applications in solar cell, photovoltaic devices, electro-optical devices, nanoscale biosensor, gas sensor, field-emission and flat-panel display, lithium ion battery, optoelectronics and photocatalysis due to the performance in terms of conductivity, transmissivity, stability and surface morphology [16-17]. Materials research focused on the synthesis and characterization of structural, electronic and optical properties of transparent conducting oxides has experienced a great interest in the past few years. Recently, Wang and his co-workers investigated the room temperature H_2S gas sensing properties of In_2O_3

micro/nanostructured porous thin film and explored the hydrolyzation-induced enhanced sensing mechanism [18]. Park and his co-workers investigated the synergistic effects of codecoration of oxide nanoparticles on the gas sensing performance of In_2O_3 nanorods [19]. He and his co-workers studied the facile synthesis of In_2O_3 nanospheres with excellent sensitivity to trace explosive nitro-compounds [20]. Ilin and his co-workers revealed the UV effect on NO_2 sensing properties of nanocrystalline In_2O_3 [21]. Gong and his co-workers investigated the porous In_2O_3 nanocuboids modified with Pd nanoparticles for chemical sensors [22]. Pantilimon and his co-workers studied the synthesis of nano-sized indium oxide (In_2O_3) powder by a polymer solution route [23]. Gong and his co-workers discussed the 3D hierarchical In_2O_3 nanoarchitectures consisting of nanocuboids and nanosheets for chemical sensors with enhanced performances [24].

Korotcenkov and his co-workers investigated the In_2O_3-based multicomponent metal oxide films and their prospects for thermoelectric applications [25]. Yang and his co-workers studied the shape-controlled synthesis and photocatalytic activity of In_2O_3 nanostructures derived from coordination polymer precursors [26]. Wang and his co-workers explored the mesoporous In_2O_3 materials prepared by solid-state thermolysis of indium-organic frameworks and their high HCHO-sensing performance [27]. Liang and his co-workers investigated the synthesis of In_2O_3 hollow nanofibers and their application in highly sensitive detection of acetone [28]. Li and his co-workers studied the vitamin C-assisted synthesis and gas sensing properties of coaxial In_2O_3 nanorod bundles [29]. Anand and his co-workers investigated the effect of terbium doping on structural, optical and gas sensing properties of In_2O_3 nanoparticles [30]. Li and his co-workers investigated the synthesis, electrochemical and gas sensing properties of In_2O_3 nanostructures with different morphologies [31]. Park and his co-workers revealed the ethanol sensing properties of networked In_2O_3 nanorods decorated with Cr_2O_3 nanoparticles [32]. Klaus and his co-workers investigated the light-activated resistive ozone sensing at room temperature utilizing nanoporous In_2O_3 particles and its influence of particle size [33]. Meng and his co-workers studied the highly sensitive and fast responsive semiconductor metal oxide detector based on In_2O_3 nanoparticle film for portable gas chromatograph [34]. Wang and his co-workers investigated the semiconducting properties of In_2O_3 nanoparticle thin films in air and nitrogen [35].

In the present study, we report on the structural properties of beryllium and indium oxide powders using a sustainable green synthesis method without surfactant and template at room temperature incubation and aging over a period of 7 days for sustainable materials development. Therefore, in the present study, metal-ellagate square planar complex method is employed to prepare metal oxide powders [36-44]. The obtained metal-ellagate complex has been decomposed into BeO and In_2O_3 powders by calcinating in static air atmosphere at 350, 500 and 650°C for 1 h.

2.   Materials and methods

2.1   Synthesis of beryllium and indium oxide powders

Initially, 0.1 M aqueous solution of beryllium nitrate hydrate $[Be(NO_3)_2.H_2O]$ and indium (III) nitrate hydrate $[In(NO_3)_3.xH_2O]$ was prepared separately. Then, 10 ml rambutan extract was added to 50 ml of 0.1 M $[Be(NO_3)_2.H_2O]$ and $[In(NO_3)_3.xH_2O]$ separately stirred at about 80°C for 2 h. The obtained product was incubated at room temperature for 7 days to form corresponding metal-ellagate complex formation and then centrifuged and air dried and then powdered using a mortar and pestle. Corresponding metal oxide powders were obtained due to direct decomposition of suitable metal-ellagate complexes in a muffle furnace at 350, 500 and 650°C for 1 h in a static air atmosphere.

2.2   Characterization of beryllium and indium oxide powders

Phase and crystalline nature of prepared samples were identified by X-ray powder diffraction (X'pert PRO analytical diffractometer) using CuKα as radiation (1.541 A°) source. Imaging Spectrograph STR 500 nm focal length laser micro Raman spectrometer SEKI, Japan with resolution: 1/0.6cm-1/pixel and Flat Field: 27 mm (W) × 14mm (H) was used. Photoluminescence (PL) spectrum was measured using a Varian Cary Eclipse Photoluminescence Spectrometer with Oxford low temperature LN277K setup. Infrared (IR) spectra were recorded using a Fourier transform infrared spectrophotometer using Thermo Nicolet 380 with a resolution of 0.5 cm-1 and S/N ratio: 2000:1 ppm for 1 minute scan.

3.   Results and discussion

3.1   XRD pattern of beryllium and indium oxide powders

The XRD pattern of the synthesized products is shown in Fig. 1(a-c). XRD analysis was performed to investigate the phases of the obtained product after calcination treatments. Fig. 1(a-c) shows the X-ray diffraction patterns from the samples calcinated at 350, 500 and 650°C in a static air atmosphere for 1 h. In all the cases, clear diffraction patterns are shown. All of these patterns agree very well with the characteristic of a hexagonal pattern. It agrees well with the JCPDS card No. 35-0818 for pure α-BeO phase, confirming the formation of a pure α-BeO at different calcination temperatures as shown in Fig. 1 (a-c). The broadened diffraction peaks indicate that the crystallite size of the sample is excellent and the diffraction intensity increased with increasing calcinations time, which means that the BeO grains have grown (Table 1). By using the data obtained from the XRD patterns and Debye–Scherrer formula: $D=0.9\lambda/(\beta\cos\theta)$, where D is the crystallite size (nm), λ is the radiation wavelength ($\lambda=0.154$ nm), θ is the diffraction peak angle, and β is the corrected half-width at half-maximum intensity (FWHM), the average crystallite sizes of samples calcinated at different temperatures such as 350, 500 and 650°C are calculated. It can be concluded that the size of the product output have a tendency to increase with the increasing the calcinations time.

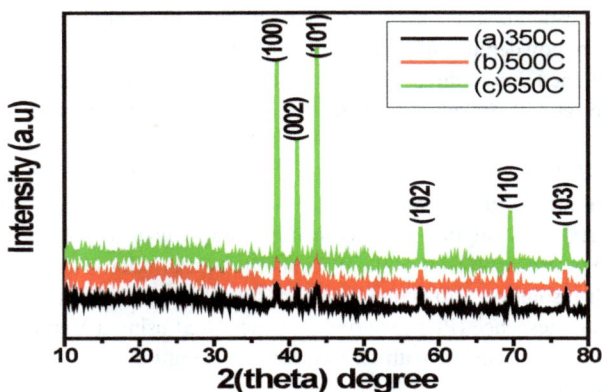

Figure 1.   XRD Pattern (BeO): 7 days of reaction product from the mixture of 0.1M beryllium nitrate and 10 ml extract calcinated at (a) 350°C, (b) 500°C and (c) 650°C.

Table 1   X-ray diffraction values from the samples calcinated at 350, 500 and 650°C.

Pos. [°2Th.]	Height [cts]	FWHM Left [°2Th.]	d-spacing [Å]	Rel. Int. [%]
X-ray diffraction values from the samples calcinated at 350°C				
38.3627	743.28	0.1968	2.34642	93.56
41.0682	496.86	0.1968	2.19787	62.54
43.7526	794.45	0.1968	2.06905	100.00
57.4949	127.41	0.2460	1.60295	16.04
69.4862	223.75	0.1476	1.35276	28.16
76.8273	146.60	0.1476	1.24078	18.45
X-ray diffraction values from the samples calcinated at 500°C				
38.3510	382.35	0.1476	2.34710	83.23
41.0531	272.89	0.1476	2.19865	59.40
43.7447	459.40	0.1968	2.06941	100.00
57.5337	68.06	0.2460	1.60196	14.82
69.4813	94.64	0.1968	1.35285	20.60
76.8046	74.83	0.1476	1.24109	16.29
X-ray diffraction values from the samples calcinated at 650°C				
38.3612	868.59	0.1968	2.34650	92.89
41.0830	532.32	0.1968	2.19712	56.93
43.7634	935.08	0.1968	2.06857	100.00
57.5046	149.93	0.1968	1.60270	16.03
69.4913	221.21	0.1476	1.35268	23.66
76.8288	156.49	0.1476	1.24076	16.74

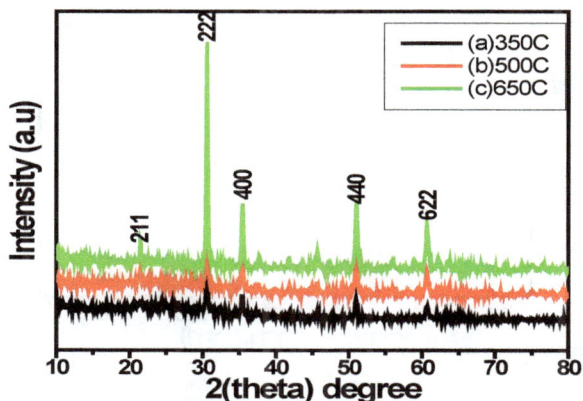

Figure 2. XRD Pattern (In_2O_3):7 days of reaction product from the mixture of 0.1M indium nitrate hydrate and 10 ml extract calcinated at (a) 350°C, (b) 500°C and (c) 650°C.

Similarly, the powder X-ray diffraction (XRD) was used to characterize the phase structure of the In_2O_3 powders calcinated at different temperature over a period of 1 h at 350°C, 500°C and 650°C. The obtained XRD pattern for In_2O_3 calcinated at three different temperatures was shown in Fig. 2a-c. In_2O_3 powders calcinated at 350°C have shown less crystalline nature at low temperature due to defect and lack of kinetic energy and mobility of the grains to get oriented in the respective plane. The observed less crystalline peaks suggest that crystallization is inhibited when the temperature is low. As the temperature increases above 350°C to 500°C, the peaks related to indium oxide crystalline phase are observed showing that the material is becoming crystalline. As the calcination temperature increases from 500°C to 650°C, the peak intensity also increases; this shows the improvement in crystallinity as the temperature increases. In_2O_3 powders calcinated at different temperature over a period of 350°C, 500°C and 650°C exhibited different crystallite size. And no other crystalline phases were detected within the detection limit. All the XRD patterns of the samples exhibit the same reflection. No other peaks can be observed revealing their phase –pure cubic structure. Indium oxide strong reflections (211), (222), (400), (440) and (622) planes with respective 2 theta value for 22, 31, 35, 51 and 61 are exactly correlated with the reported values. All the XRD reflection can be indexed to cubic In_2O_3 (JCPDS, No 65-3170).

3.2 PL spectra of beryllium and indium oxide powders

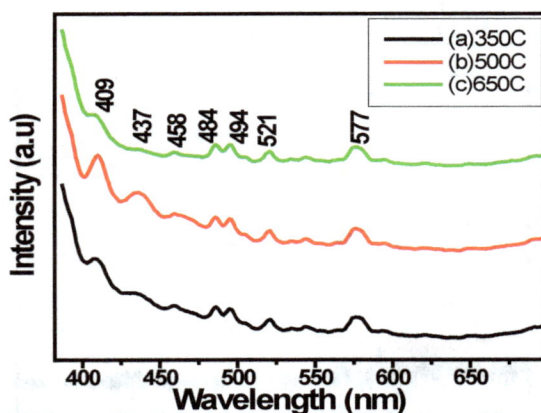

Figure 3.   Room-temperature PL spectra: 7 days of reaction product from the mixture of 0.1M beryllium nitrate and 10 ml extract calcinated at (a) 350°C, (b) 500°C and (c) 650°C.

Fig. 3 exhibits a room temperature Photo Luminescence (PL) spectrum of BeO powders incubated over 7 days and then calcinated at different temperature over a period of 1 h at 350°C, 500°C and 650°C. The excitation was conducted under 250 nm UV light from Xe lamp. It is revealed that the PL peaks observed at 409, 437, 458, 484, 494, 521, 577 nm are exactly correlated to the reported values.

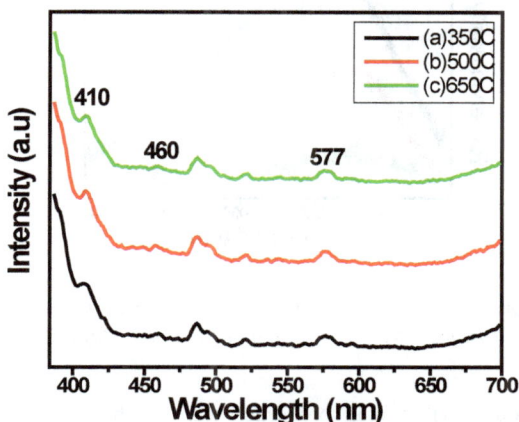

Figure 4. Room-temperature PL spectra:7 days of reaction product from the mixture of 0.1M indium nitrate hydrate and 10 ml extract calcinated at (a) 350°C, (b) 500°C and (c) 650°C.

Fig. 4a-c exhibits a room temperature Photo Luminescence (PL) spectrum of In_2O_3 powders incubated over 7 days and then calcinated at different temperature over a period of 1h at 350°C, 500°C and 650°C. The excitation was conducted under 250 nm ultraviolet light from Xe lamp. PL emission was mainly attributed to reveal the presence of indium or oxygen vacancies or to reveal the defects due to the impurities or interstitial sample. Room-temperature PL spectra of In_2O_3 powders incubated at different temperature over a period of 1h at 350°C, 500°C and 650°C clearly revealed that the blue emission was observed at 410 nm for all the samples. This blue emission can be attributed to the presence of oxygen vacancies created during the calcinations process. Nearly identical strong PL peaks centered at 460 nm (blue) and 577 nm (green) were exactly correlated to the reported values of cubic In_2O_3.

3.3   Raman spectra of beryllium and indium oxide powders

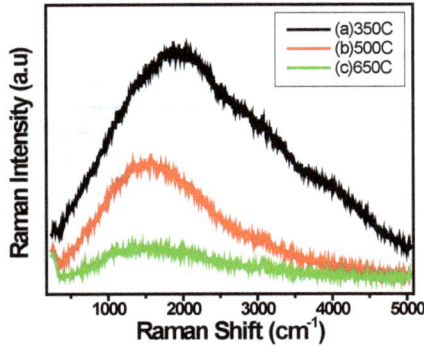

Figure 5.   Raman Spectra: 7 days of reaction product from the mixture of 0.1M beryllium nitrate and 10 ml extract calcinated at (a) 350°C, (b) 500°C and (c) 650°C.

The obtained product's vibrational properties were further studied employing micro Raman spectroscopy. Fig. 5(a-c) shows the micro Raman spectra of BeO powders incubated over 7 days and then calcinated at various temperatures namely 350°C, 500°C and 650°C. The observed strong sharp band around 1500 cm^{-1} was clearly revealed the pure α-BeO phase and confirmed the Raman vibration modes of α-BeO.

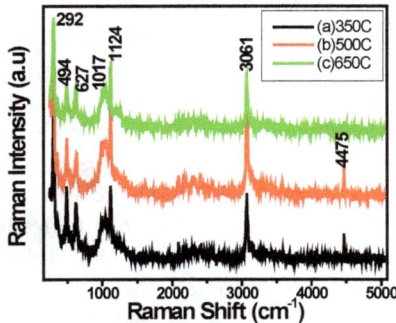

Figure 6.   Raman Spectra: 7 days of reaction product from the mixture of 0.1M indium nitrate hydrate and 10 ml extract calcinated at (a) 350°C, (b) 500°C and (c) 650°C.

The defect states of In_2O_3 powders calcinated at different temperature over a period of 1h at 350°C, 500°C and 650°C were analyzed using room temperature micro Raman-scattering as shown in Fig. 6a-c. Fig. 6a-c displays the Raman shift data from 200–5000 cm^{-1}. The observed frequency sets at 292 (bending vibration), 494 and 627 cm^{-1} belong to the stretching vibration modes of bcc-In_2O_3, which agree well with the reported values in the literature. Fig. 6(a-c) revealed the micro-Raman spectra of the product calcinated at (a) 350°C, (b) 500°C and (c) 650°C and confirmed the Raman vibration modes of bcc-In_2O_3.

3.4   IR spectra of beryllium and indium oxide powders

The obtained IR broad absorption peaks around 3449 cm^{-1} were due to O-H stretching of absorbed H_2O molecules (Fig. 7a-c). The obtained IR peaks around 1628 cm^{-1} and 1404 cm^{-1} bands were most likely due to carbonate moieties. Additionally, the appearance of a broad band in the range of 599 cm^{-1} confirmed the characteristic band assigned to Be-O stretching mode. The obtained characteristic peaks were evidenced the BeO formation.

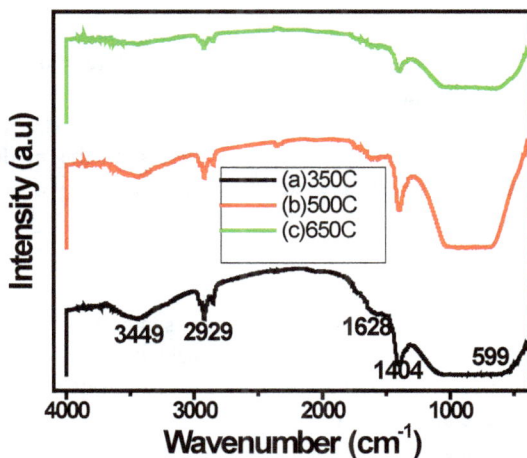

Figure 7.   IR Spectra: 7 days of reaction product from the mixture of 0.1M beryllium nitrate and 10 ml extract calcinated at (a) 350°C, (b) 500°C and (c) 650°C.

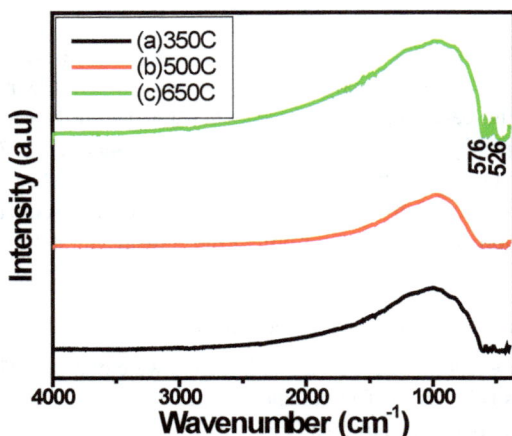

Figure 8.   IR Spectra:7 days of reaction product from the mixture of 0.1M indium nitrate hydrate and 10 ml extract calcinated at (a) 350°C, (b) 500°C and (c) 650°C.

The FTIR spectra of 7 days of reaction product from the mixture of 0.1M indium nitrate and 10 ml extract calcinated at (a) 350°C, (b) 500°C and (c) 650°C are shown in Fig. 8a-c. The spectra were recorded in the range of 500–4000 cm^{-1}. The absorption band observed at 576 and 526 cm^{-1} were associated with In–O stretching modes of vibration. The appearance of this band clearly revealed the presence of In$_2$O$_3$ in the crystalline phase. No other band was observed.

Conclusions

In summary, a simple method employing a green synthesis technique to prepare beryllium and indium oxide powder over a period of incubation at room temperature was explored to study the structural properties of the obtained powders. We demonstrate a green synthesis method employing corresponding metal-ellagate complex formation method. The effect of incubation and calcination temperature on the structural properties of the obtained product was investigated. The results indicate that the proper incubation and aging is an effective tool to prepare structured metal oxide powders.

References

[1] P. J. Anderson and R. F. Horlock, Calcination of microporous BeO powders, Trans. Faraday Soc. 63 (1967) 2316-2323. http://dx.doi.org/10.1039/tf9676302316

[2] V.S. Kiiko, I.A. Dmitriev, Y.N. Makurin, A.A. Sofronov, A.L. Ivanovskii, Synthesis and Application of Transparent Beryllium Ceramics, Glass Phys. Chem 30 (2004) 109-111. http://dx.doi.org/10.1023/B:GPAC.0000016407.00973.8c

[3] T.J. Oatts, C.E. Hicks, A.R. Adams, M.J. Brisson, L.D. Youmans-McDonald, M.D. Hoover, K. Ashley, Preparation, certification and interlaboratory analysis of workplace air filters spiked with high-fired beryllium oxide, J. Environ. Monit. 14 (2012) 391-401. http://dx.doi.org/10.1039/C1EM10688K

[4] M.D. Petrenko, I.N. Ogorodnikov, V. Yu. Ivanov, Thermoluminescence and low-temperature luminescence of beryllium oxide, Radiat. Meas. doi:10.1016/j.radmeas.2015.12.025. http://dx.doi.org/10.1016/j.radmeas.2015.12.025

[5] S. Dawahra, K. Khattab, G. Saba, Investigation of BeO as a reflector for the low power research reactor, Prog. Nucl. Energy 81 (2015) 1-5. http://dx.doi.org/10.1016/j.pnucene.2014.12.001

[6] A. Fathalian, F. Kanjouri, J. Jalilian, BeO nanotube bundle as a gas sensor, Superlattices Microstruct. 60 (2013) 291-299. http://dx.doi.org/10.1016/j.spmi.2013.04.028

[7] Fathalian, R. Moradian, M. Shahrokhi, Optical properties of BeO nanotubes: Ab initio study, Solid State Commun. 156 (2013) 1–7. http://dx.doi.org/10.1016/j.ssc.2012.11.017

[8] J. Baima, A. Erba, M. Rerat, R. Orlando, R. Dovesi, Beryllium Oxide Nanotubes and their Connection to the Flat Monolayer, J. Phys. Chem. C, 117 (2013) 12864–12872. http://dx.doi.org/10.1021/jp402340z

[9] E.C. Anota, G.H. Cocoletzi, Electronic properties of functionalized (5,5) beryllium oxide nanotubes, J. Mol. Graphics Modell. 42 (2013) 115-119. http://dx.doi.org/10.1016/j.jmgm.2013.03.007

[10] E.O. Wrasse, R.J. Baierle, First principles study of native defects in BeO, Physics Procedia 28 (2012) 79-83. http://dx.doi.org/10.1016/j.phpro.2012.03.675

[11] W. Wu, P. Lu, Z. Zhang, W. Guo, Electronic and magnetic properties and structural stability of BeO sheet and nanoribbons, ACS Appl. Mater. Interfaces, 3 (2011) 4787–4795. http://dx.doi.org/10.1021/am201271j

[12] X. Wang, R. Wang, C. Peng, T. Li, B. Liu, Growth of BeO Nanograins Synthesized by Polyacrylamide Gel Route, J. Mater. Sci. Technol. 7 (2011) 147-152. http://dx.doi.org/10.1016/S1005-0302(11)60040-6

[13] X. Wang, R. Wang, C. Peng, T. Li, B. Liu, Synthesis and sintering of beryllium oxide nanoparticles, Prog. Nat. Sci. 20 (2010) 81–86. http://dx.doi.org/10.1016/S1002-0071(12)60011-2

[14] X. Sun, H. Hao, H. Ji, X. Li, S. Cai, C. Zheng, Synthesis of In_2O_3 with Appropriate Mesostructured Ordering and Enhanced Gas-Sensing Property, ACS Appl. Mater. Interfaces 6 (2014) 401–409. http://dx.doi.org/10.1021/am4044807

[15] L. Yin, D. Chen, M. Hu, H. Shi, D. Yang, B. Fan, G. Shao, R. Zhang, G. Shao, Microwave-assisted growth of In_2O_3 nanoparticles on WO_3 nanoplates to improve H_2S-sensing performance, J. Mater. Chem. A2 (2014) 18867-18874. http://dx.doi.org/10.1039/C4TA03426K

[16] G. Wang, J. Park, D. Wexler, M.S. Park, J.H. Ahn, Synthesis, Characterization, and Optical Properties of In_2O_3 Semiconductor Nanowires, Inorg. Chem. 46 (2007) 4778–4780. http://dx.doi.org/10.1021/ic700386z

[17] J.S. Lee, Y.J. Kwack, W.S. Choi, Inkjet-Printed In_2O_3 Thin-Film Transistor below 200°C, ACS Appl. Mater. Interfaces 5 (2013) 11578–11583. http://dx.doi.org/10.1021/am4025774

[18] Y. Wang, G. Duan, Y. Zhu, H. Zhang, Z. Xu, Z. Dai, W. Cai, Room temperature H_2S gas sensing properties of In_2O_3 micro/nanostructured porous thin film and hydrolyzation-induced enhanced sensing mechanism, Sens. Actuators B 228 (2016) 74-84. http://dx.doi.org/10.1016/j.snb.2016.01.002

[19] S. Park, G.J. Sun, H. Kheel, W.I. Lee, S. Lee, S.B. Choi, C. Lee, Synergistic effects of codecoration of oxide nanoparticles on the gas sensing performance of In_2O_3 nanorods, Sens. Actuators B 227 (2016) 591-599. http://dx.doi.org/10.1016/j.snb.2015.12.098

[20] Y.Y. He, X. Zhao, Y. Cao, X. Zou, G.D. Li, Facile synthesis of In_2O_3 nanospheres with excellent sensitivity to trace explosive nitro-compounds, Sens. Actuators B 228 (2016) 295-301. http://dx.doi.org/10.1016/j.snb.2016.01.042

[21] A. Ilin, M. Martyshov, E. Forsh, P. Forsh, M. Rumyantseva, A. Abakumov, A. Gaskov, P. Kashkarov, UV effect on NO_2 sensing properties of nanocrystalline In_2O_3, Sens. Actuators B Available online 16 March 2016.

[22] F. Gong, Y. Gong, H. Liu, M. Zhang, Y. Zhang, F. Li, Porous In_2O_3 nanocuboids modified with Pd nanoparticles for chemical sensors, Sens. Actuators B 223 (2016) 384-391. http://dx.doi.org/10.1016/j.snb.2015.09.053

[23] M.C. Pantilimon, T.S. Kang, S.J. Lee, Synthesis of nano-sized indium oxide (In_2O_3) powder by a polymer solution route, Ceram. Int. 42 (2016) 3762-3768. http://dx.doi.org/10.1016/j.ceramint.2015.11.006

[24] F. Gong, H. Liu, C. Liu, Y. Gong, Y. Zhang, E. Meng, F. Li, 3D hierarchical In_2O_3 nanoarchitectures consisting of nanocuboids and nanosheets for chemical sensors with enhanced performances, Mat. Lett. 163 (2016) 236-239. http://dx.doi.org/10.1016/j.matlet.2015.10.106

[25] G. Korotcenkov, V. Brinzari, B.K. Cho, In_2O_3-based multicomponent metal oxide films and their prospects for thermoelectric applications, Solid State Sci. 52 (2016) 141-148. http://dx.doi.org/10.1016/j.solidstatesciences.2015.12.019

[26] J.M. Yang, Z.P. Qi, Y.S. Kang, Q. Liu, W.Y. Sun, Shape-controlled synthesis and photocatalytic activity of In_2O_3 nanostructures derived from coordination polymer precursors, Chin. Chem. Lett.   Available online 14 January 2016.

[27] J.L. Wang, Q.G. Zhai, S.N. Li, Y.C. Jiang, M.C. Hu, Mesoporous In_2O_3 materials prepared by solid-state thermolysis of indium-organic frameworks and their high HCHO-sensing performance, Inorg. Chem. Commun. 63 (2016) 48-52. http://dx.doi.org/10.1016/j.inoche.2015.11.015

[28] X. Liang, G. Jin, F. Liu, X. Zhang, S. An, J. Ma, G. Lu, Synthesis of In_2O_3 hollow nanofibers and their application in highly sensitive detection of acetone, Ceram. Int. 41 (2015) 13780-13787. http://dx.doi.org/10.1016/j.ceramint.2015.08.060

[29] X. Li, S. Yao, J. Liu, P. Sun, Y. Sun, Y. Gao, G. Lu, Vitamin C-assisted synthesis and gas sensing properties of coaxial In_2O_3 nanorod bundles, Sens. Actuators B 220 (2015) 68-74. http://dx.doi.org/10.1016/j.snb.2015.05.038

[30] K. Anand, J. Kaur, R.C. Singh, R. Thangaraj, Effect of terbium doping on structural, optical and gas sensing properties of In_2O_3 nanoparticles, Mater. Sci. Semicond. Process. 39 (2015) 476-483. http://dx.doi.org/10.1016/j.mssp.2015.05.042

[31] F. Li, J. Jian, R. Wu, J. Li, Y. Sun, Synthesis, electrochemical and gas sensing properties of In_2O_3 nanostructures with different morphologies, J. Alloys Compd. 645 (2015) 178-183. http://dx.doi.org/10.1016/j.jallcom.2015.04.157

[32] S. Park, S. Kim, G.J. Sun, S. Choi, S. Lee, C. Lee, Ethanol sensing properties of networked In_2O_3 nanorods decorated with Cr_2O_3 nanoparticles, Ceram. Int. 41 (2015) 9823-9827. http://dx.doi.org/10.1016/j.ceramint.2015.04.055

[33] D. Klaus, D. Klawinski, S. Amrehn, M. Tiemann, T. Wagner, Light-activated resistive ozone sensing at room temperature utilizing nanoporous In_2O_3 particles: Influence of particle size, Sens. Actuators B 217 (2015) 181-185. http://dx.doi.org/10.1016/j.snb.2014.09.021

[34] H. Meng, W. Yang, X. Yan, Y. Zhang, L. Feng, Y. Guan, A highly sensitive and fast responsive semiconductor metal oxide detector based on In_2O_3 nanoparticle film for portable gas chromatograph, Sens. Actuators B 216 (2015) 511-517. http://dx.doi.org/10.1016/j.snb.2015.04.068

[35] L. Wang, X. Xu, Semiconducting properties of In_2O_3 nanoparticle thin films in air and nitrogen, Ceram. Int. 41 (2015) 7687-7692. http://dx.doi.org/10.1016/j.ceramint.2015.02.097

[36] R. Yuvakkumar, J. Suresh, A. Joseph Nathanael, M. Sundrarajan, S.I. Hong, Novel green synthetic strategy to prepare ZnO nanocrystals using rambutan (Nephelium lappaceum L.) peel extract and its antibacterial applications, Mater. Sci. Eng. C 41 (2014) 17-27. http://dx.doi.org/10.1016/j.msec.2014.04.025

[37] R. Yuvakkumar, J. Suresh, A. Joseph Nathanael, M. Sundrarajan, S.I. Hong, Rambutan (Nephelium lappaceum L.) peel extract assisted biomimetic synthesis of nickel oxide nanocrystals, Mat. Lett. 128 (2014) 170-174. http://dx.doi.org/10.1016/j.matlet.2014.04.112

[38] R. Yuvakkumar, J. Suresh, B. Saravanakumar, A. Joseph Nathanael, S.I. Hong, V. Rajendran, Rambutan peels promoted biomimetic synthesis of bioinspired zinc oxide nanochains for biomedical applications, Spectrochim. Acta Part A. 137 (2015) 250-258. http://dx.doi.org/10.1016/j.saa.2014.08.022

[39] R. Yuvakkumar, A. Joseph Nathanael, S.I. Hong, Inorganic complex intermediate Co_3O_4 nanostructures using green ligation from natural waste resources, RSC Adv. 4 (2014) 44495–44499. http://dx.doi.org/10.1039/C4RA07646J

[40] R. Yuvakkumar, S.I. Hong, Incubation and aging effect on cassiterite type tetragonal rutile SnO_2 nanocrystals, J. Mater. Sci. - Mater. Electron. 26 (2015) 2305-2310. http://dx.doi.org/10.1007/s10854-015-2684-1

[41] R. Yuvakkumar, J. Suresh, B. Saravanakumar, A. Joseph Nathanael, V. Rajendran, S.I. Hong, An environment benign biomimetic synthesis of mesoporous NiO

concentric stacked doughnuts architecture, Microporous Mesoporous Mater. 207 (2015) 185–194. http://dx.doi.org/10.1016/j.micromeso.2015.01.027

[42] R. Yuvakkumar, A. Joseph Nathanael, S.I. Hong, Nd_2O_3: Novel synthesis and characterization, J. Sol-Gel Sci. Technol. 73 (2015) 511-517. http://dx.doi.org/10.1007/s10971-015-3629-0

[43] R. Yuvakkumar, S.I. Hong, Structural, compositional and textural properties of monoclinic α-Bi_2O_3 nanocrystals, Spectrochim. Acta Part A. 144 (2015) 281-286. http://dx.doi.org/10.1016/j.saa.2015.02.093

[44] R. Yuvakkumar, S.I. Hong, Structural phase transitions in niobium oxide nanocrystals, Phase Transitions 88 (2015) 897-906.

CHAPTER 10

Charge density analysis and magnetic behavior of Li doped NiO nanostructures synthesized by sol-gel process

K. Sakthi Lavanya[1], B. Subha[2], M. Prema Rani*[1], R.Saravanan[1]

[1]Research Centre and PG Department of Physics, The Madura College, Madurai-11, Tamil Nadu, India

[2]PG Department of Physics, E.M.G. Yadava Women's college, Madurai-14, Tamil Nadu, India

Email: premaakumar@yahoo.com

Abstract

Li-doped nickel oxide nanostructures ($Ni_{1-x}Li_xO$, x=0, 0.03 and 0.06) of crystallite size of around 20 nm have been prepared using the sol-gel process. The cell constant of the prepared samples decreases with the concentration of Li dopant. This ensures that Li occupies the host lattice of Ni. The electronic charge distributions in the unit cell were analyzed through the maximum entropy method (MEM) for the prepared cubic nickel nanostructure. The bonding features of the prepared doped nanostructures were analyzed and they are found to behave like covalent materials. From the UV analysis, the band gap was determined as 3.28 eV for NiO. Strong ferromagnetic behavior is observed for 3% Li doped NiO. Further addition of Li dopant causes decrease in ferromagnetism.

Keywords

Nanostructures, XRD, Rietveld Refinement, MEM, Electron Density

Contents

1.   Introduction

Over recent years, increasing attention has been given to the production of novel nanostructured metal oxide materials. One of the most commonly used transition metal oxides for a wide range of applications is NiO. Nickel oxide may exist in various forms like NiO, NiO_2, NiO_4 and Ni_2O_3. Among them, NiO is an anti-ferromagnetic material with a density of 6.67 g/cm^3 and having a cubic structure with wide band-gap energy ranging from 3.6 to 4.0 eV [1]. Interest in the magnetic properties of nanosized NiO has recently been revived due to reports of interesting and complex magnetic behavior [2, 3]. Although bulk NiO is antiferromagnetic, nanoparticles exhibit room-temperature coercivity, spin-glass behavior, superparamagnetism, memory effects, exchange bias effects and magnetic transitions at low temperature [4, 5]. Nickel oxide (NiO) is a promising material, extensively used in catalysis, battery cathodes, gas sensors and electrochromic films, photo-electronic devices, [6,7]. NiO has a wide range of applications, such as transparent conductive films , electrochromic display devices, anode material in organic light emitting diodes, and functional layer material for chemical sensors, optical gas sensing, environmental pollution monitoring and chemical process control [8-14]. Lately, there has been a lot of interest in nickel oxide's optical and electrochromic properties connected with its possible applications in displays and "smart windows". Especially semiconducting NiO would be usefully applied in UV detection [15]. Several methods have been used and developed for synthesizing crystalline oxide powders in nanoscale dimensions. Many researchers have synthesized NiO nanoparticles by various methods, such as sol–gel, surfactant-mediated synthesis, thermal decomposition, and solvothermal process [16-20].

2.   Experimental procedure

$Ni_{1-x}Li_xO$ (x = 0, 0.03, 0.06) nanostructures have been prepared through the sol-gel method combined with sintering. Citric acid was dissolved in ethanol with stirring at room temperature. Then, suitable quantities of $Ni(CH_3COO)_2.4H_2O$ and $CH_3COOLi.H_2O$ (Ni:Li=100:0;100:3;100:6) were added into the solution under vigorous stirring at ambient temperature till a viscous solution was obtained. The solution was heated until a green gel was obtained. Then, the gel was dried at 80°C for 5 h. A green dry gel was obtained. After calcining at 485°C for 2 h in air, pure NiO powder and Li-doped NiO powders were obtained. The prepared samples were characterized by a X-ray diffractometer (XRD), Scanning electron microcopy (SEM), Ultra Violet spectroscopy (UV) and Vibrating Sample Magnetometer (VSM) for structural, morphological, optical and magnetic studies respectively.

3.   Results and discussion

3.1   X-ray analysis

The X-ray characterization was done for the prepared powder samples of $Ni_{1-x}Li_xO$ (x = 0, 0.03 and 0.06) nanostructures using an X'Pert PRO (Philips, Netherlands) X-ray diffractometer with CuK_α monochromator in the range of 5° to120° with the step size of 0.02°. The collected X-ray profiles for different concentration of Li doped NiO nanostructures are shown in Fig. 1. The observed X-ray peaks for the prepared nanostructures were matched with standard pattern of (JCPDS) XRD data set reported in the file (pdf No: 471049). Li stabilized cubic nickel nanostructures were identified with the space group of Fm3m. The profiles without any additional peaks show that the dopant atoms occupy the preferential sites in the host lattice sites. Fig. 2 shows the shift in the diffracting angle 2θ towards higher angles from XRD profile peak of (002) Bragg reflection. This shift in (002) peak position towards higher 2θ may be due to the contraction of lattice and hence the lattice parameters. Since the inter planar distance and 2θ are inversely proportional, the cell constant decreases with the concentration of Li-dopant.

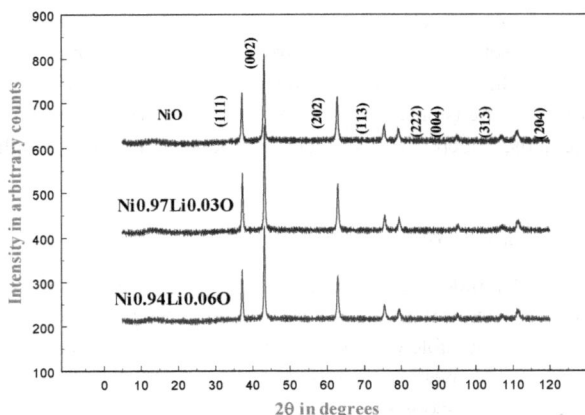

Figure 1. XRD profiles of $Ni_{1-x}Li_xO$ nanostructures.

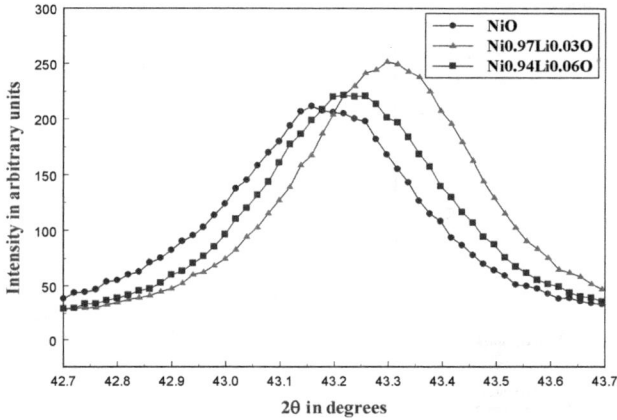

Figure 2. Shift in the diffracting angle 2θ towards higher angles for (002) for $Ni_{1-x}Li_xO$ nanostructures.

The effect of Li concentration on the structural features on NiO nanostructures has been analyzed using the Rietveld refinement [21] which is employed in the software JANA 2006 [22]. It has been used to match the experimental and calculated diffraction patterns by considering the cubic structure which belongs to the space group Fm3m with four molecules in the unit cell. In Rietveld refinement technique the difference between the theoretically modelled profile and the observed one is minimised. The refined powder profiles of $Ni_{1-x}Li_xO$ (x =0, 0.03 and 0.06) nanostructures are shown in Fig. 3(a-c). The refined cell parameters and structural parameters are given in Table 1. The cell value decreases with the increase of Li content which may be due to the higher ionic radius of Li^+ (r_{Li} = 0.76 Å) compared to Ni^{2+} (r_{Ni} = 0.69 Å). Debye Waller value (B) of Nickel increases with the increase in concentration of Li-doping. This ensures that Li occupies host lattice of Ni.

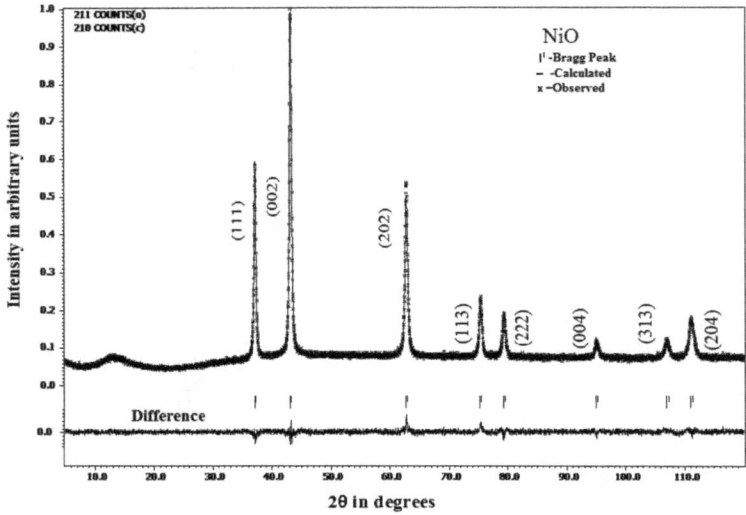

Figure 3a. Rietveld refined powder profiles of $Ni_{1-x}Li_xO$ (x =0) nanostructures.

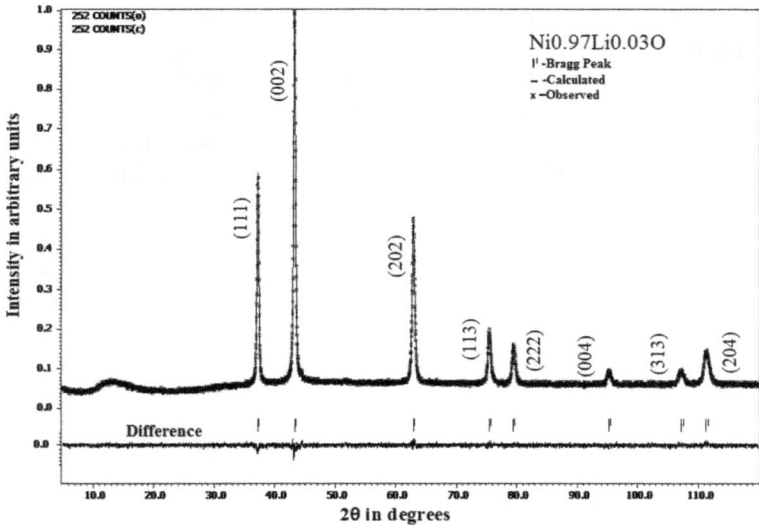

Figure 3b. Rietveld refined powder profiles of $Ni_{1-x}Li_xO$ (x = 0.03) nanostructures.

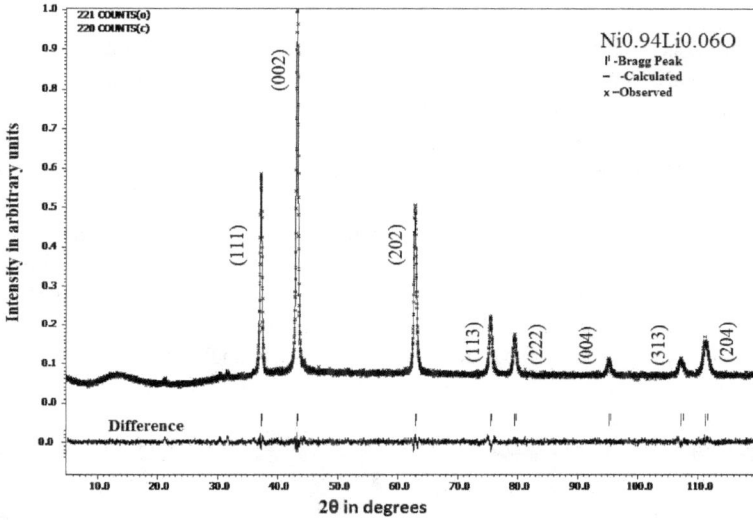

Figure 3c. *Rietveld refined powder profiles of $Ni_{1-x}Li_xO$ (x = 0.06) nanostructures.*

Table 1. Structural parameters of $Ni_{1-x}Li_xO$ nanostructures.

Parameters	x=0	x=0.03	x=0.06
Cell Parameters a=b=c(Å)	4.1757(11)	4.1718(32)	4.1706(05)
α=β=γ	90°	90°	90°
Volume(Å³)	72.8094	72.6075	72.5405
Electrons in the unit cell	144	141	138
Debye Waller factor(Å²)	0.855028	1.013342	1.050268
Reliable index R_{obs}(%)	0.80	1.71	1.07
Profile reliable index (Rp%)	3.28	3.48	3.71

3.2   Surface morphological analysis

The scanning electron microscope (SEM) characterization has been done for the prepared $Ni_{1-x}Li_xO$ nanostructures with the magnification of 5,000. The surface morphology of $Ni_{1-x}Li_xO$ nanostructures (x = 0, 0.03 and 0.06) are shown in Fig. 4 (a-c). From the SEM images, the average particle size (r_{SEM}) of the prepared samples was determined as 359,

229, 250 nm for x = 0, 0.03, 0.06 respectively. The crystallite sizes of the prepared samples were also found using prominent X-ray peaks using Scherrer's formula $D = k \lambda/\beta \cos\theta$ where D is the crystallite size, λ is the wavelength of X-ray (1.54056 A°) for CuKα radiation, β is the full width at half maximum in radian for the prominent intensity peak, k is a constant usually taken as 0.89 for a spherical sample of cubic symmetry. The average crystallite size (r_{Xray}) of the prepared nanostructures was determined as 19.68 nm. The observed EDAX spectra are shown for the prepared $Ni_{1-x}Li_xO$ nanostructures in Fig. 5(a-c). The peaks in the spectra indicate presence of relevant elements. The values of each element present in the prepared nanostructures are tabulated and are shown in Table 2. From Table 2, it is observed that there is decrease in weight % of Ni atom as Li dopant increases. And it also shows that reduction in weight percent due to addition of lithium. Due to low energy level of lithium (third element in periodic table) atom the corresponding peak is not shown in figure 5 as the detectable range starts from fourth element in periodic table with energy level of 0.03 keV.

Figure 4a.   *Figure 4b.*

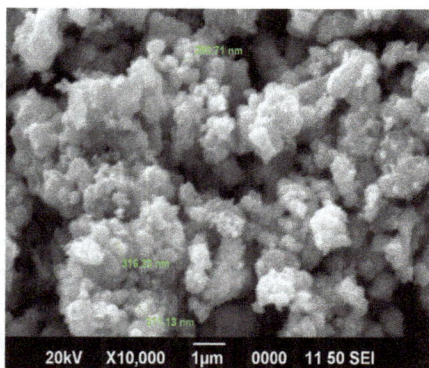

Figure 4c.

Figure 4(a-c). Scanning Electron Microscope image of $Ni1_{-x}Li_xO(x = 0, 0.03, 0.06$ respectively) nanostructures

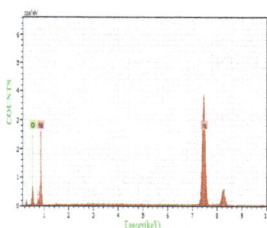 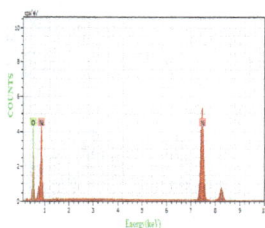 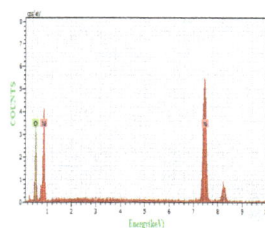

| *Figure 5a* | *Figure 5b* | *Figure 5c* |

Figures 5(a-c). $Ni1_{-x}Li_xO(x = 0, 0.03, 0.06$ respectively) nanostructures with their elemental composition found using EDAX

Table 2. Elemental composition from EDAX of $Ni_{1-x}Li_xO$ nanostructures.

Elemental	NiO		$Ni_{0.97}Li_{0.03}O$		$Ni_{0.94}Li_{0.06}O$	
	Weight%	Atom%	Weight%	Atom%	Weight%	Atom%
Ni	86.92	57.43	69.31	33.50	57.59	30.61
O	17.56	42.57	33.22	66.50	35.59	69.39
Total	104.48	100.00	94.61	100.00	93.18	100.00

3.3   Charge density analysis

Maximum entropy method (MEM) [23] is an important and accurate technique to deal the electron density distribution in the unit cell because of their probabilistic approach. Also, it only needs minimum number of information from the observed XRD spectra and it yields least biased information. This method is packaged by the software PRactice Iterative MEM Analyses (PRIMA) [24]. The structure factors extracted from Rietveld refinement technique were used for this study. The electron density distribution in the unit cell was constructed through the PRIMA software. The results are visualized by the visualization software VESTA. Three dimensional electron density distributions of $Ni_{1-x}Li_xO$ nanostructures is drawn and shown in Fig. 6(a-c) using an isosurface level of 0.6 $e/Å^3$. The position of Ni and O atoms are clearly located through the charge distribution. The shaded region surrounded by the electron cloud shows an atom. Two dimensional electron density distributions for Ni and O atom are drawn in the range of $0 - 1.0 \, e/Å^3$ with the interval of $0.1 e/Å^3$ and are shown in Fig. 7(a-c). This figure shows Ni and O atoms on the (110) miller plane. The variations in the charge density distribution for various doping concentrations is clearly visualised in the Fig. 7(a-c). The electron density distribution between the Ni and O atoms is shown in Fig. 8(a-c), as one dimensional electron density profile. The values of BCP (Bond Critical Point) between Ni and O atoms are given in Table 3.

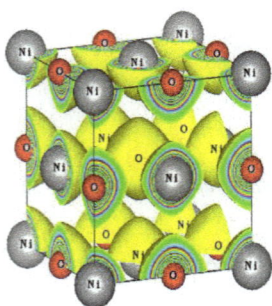 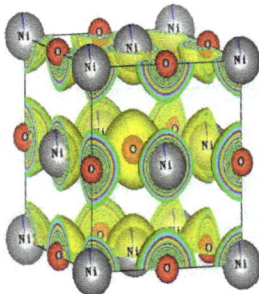 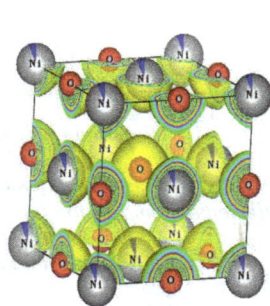

Figure 6(a).   *Figure 6(b).*   *Figure 6(c).*

Figures 6(a-c). Three dimensional electron density distribution in the unit cell (isosurface level = 0.6 e/Å³) for $Ni_{1-x}Li_xO$ (x = 0, 0.03, 0.06 respectively) nanostructures.

| Figure 7(a). | Figure 7(b). | Figure 7(c). |

Figures 7(a-c). Two dimensional electron density distribution in the unit cell in the contour range of 0 to 1.0 $e/\AA^3$ with contour interval of 0.055 $e/\AA^3$ for $Ni_{1-x}Li_xO$(x = 0, 0.03, 0.06 respectively) nanostructures.

Covalent nature of the bond is clearly visualized from the contours of Fig. 7(a-c). The quantitative data of the charge from Table 3 shows an enhanced distribution for $Ni_{0.97}Li_{0.03}O$ along [110 direction].

Table 3. BCP (Bond Critical Point) at Ni–O and Ni-Ni bond of $Ni_{1-x}Li_xO$ nanostructures.

Sample	[100] direction		[110] direction		[111] direction	
	Distance(Å)	Electron Density ($e/\AA^3$)	Distance(Å)	Electron Density ($e/\AA^3$)	Distance(Å)	Electron Density ($e/\AA^3$)
NiO	0.9862	0.6282	1.4838	0.2564	1.8172	0.1378
$Ni_{0.97}Li_{0.03}O$	0.9853	0.6423	1.4824	0.2616	1.8155	0.1212
$Ni_{0.94}Li_{0.06}O$	0.9850	0.6306	1.4819	0.2420	1.8150	0.1344

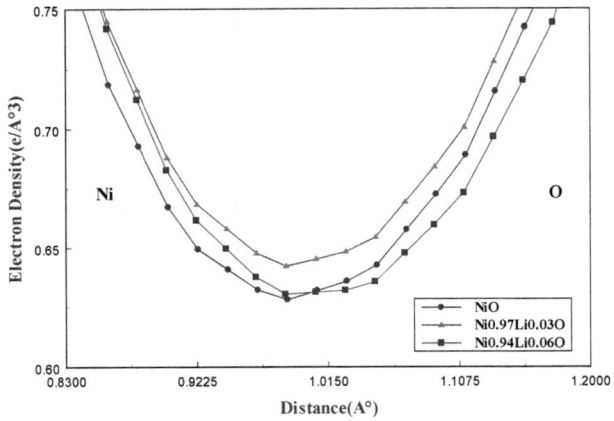

Figure 8a. One dimensional electron density profiles along [100] direction for Ni$_{1-x}$Li$_x$O nanostructures between Ni and Oxygen atoms.

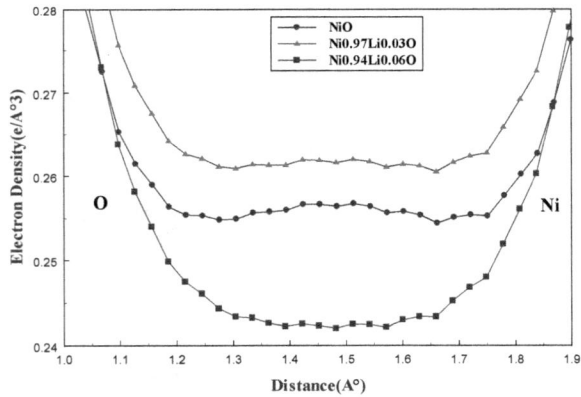

Figure 8b. One dimensional electron density profiles along [110] direction for Ni$_{1-x}$Li$_x$O nanostructures between Ni and Oxygen atoms.

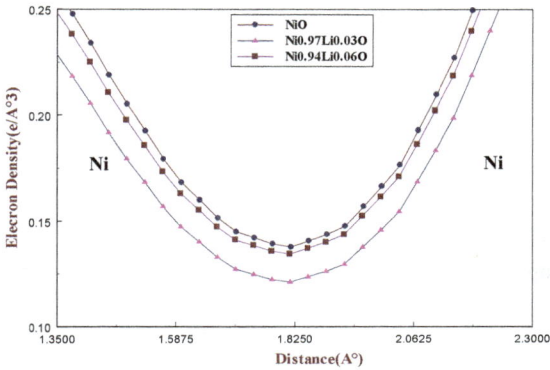

Figure 8c. One dimensional electron density profiles along [111] direction for $Ni_{1-x}Li_xO$ nanostructures between Ni and Oxygen atoms.

3.4   Analysis from ultra-violet specroscopy

UV analysis of the sample has been done at Sophisticated Analytical Instrument Facility (SAIF), Cochin, in the wavelength range of 200 to 800 nm. The UV spectrums are shown in Fig.9 for NiO and $Ni_{1-x}Li_xO$ nano structures. The graph is plotted between (hv) vs $(\alpha h v)^2 (eV/cm)^2$. Where, α is the absorption coefficient, h is the Plank constant, and v is the frequency. Calculated band gap from UV analysis are tabulated in Table 4.

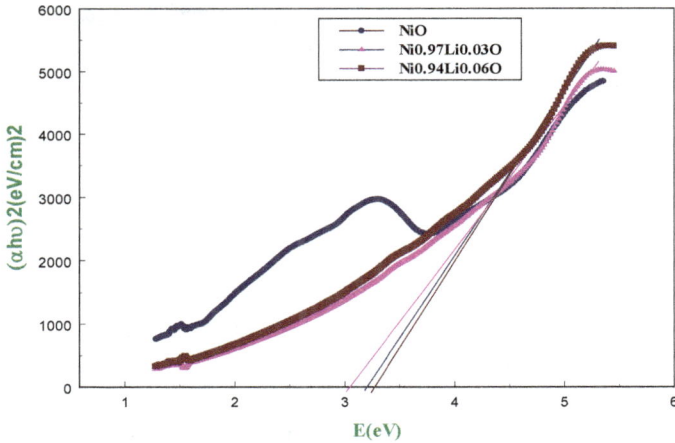

Figure 9. UV spectrum of $Ni_{1-x}Li_xO$ nanostructures.

Table 4. Band gap from UV data.

Sample	Band Gap (eV)
NiO	3.28
$Ni_{0.97}Li_{0.03}O$	3.23
$Ni_{0.94}Li_{0.06}O$	3.34

3.5   Magnetic properties

The vibrating sample magnetometer measurement has been done to measure the magnetic behavior of the prepared nanostructures of $Ni_{1-x}Li_xO$ using the vibrating sample magnetometer of Lakeshore VSM 7410 that works on Faraday's law of induction. Constant magnetic field is given to the sample and it was allowed to vibrate. The magnetic field was varied over a given range and a plot of magnetization (M) versus magnetic field strength (H) was generated. The resultant graph is shown in Fig. 10. The observed values are also tabulated in Table 5.

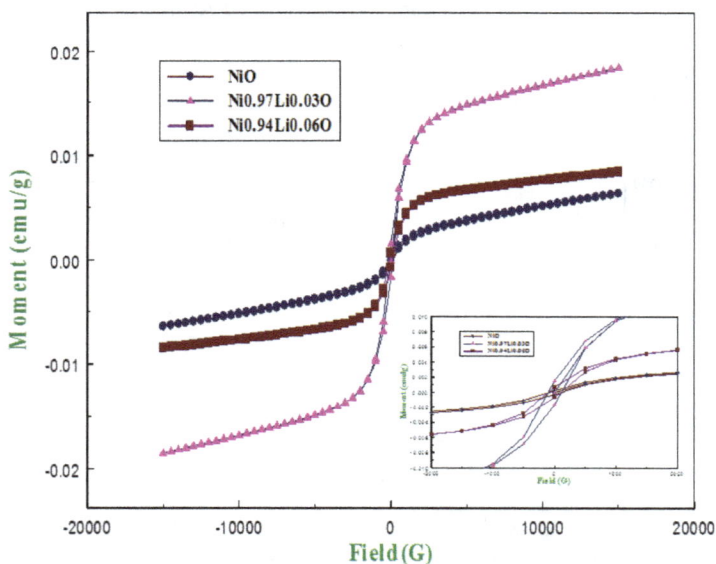

Figure 10. VSM Spectrum of $Ni_{1-x}Li_xO$ nanostructures

Table 5. Magnetic parameters from VSM measurements

Parameter	NiO	$Ni_{0.97}Li_{0.03}O$	$Ni_{0.94}Li_{0.06}O$
Magnetization(M_s) (memu)	6.3815	18.476	8.4295
Coercivity(H_{ci}) (G)	103.32	106.44	100.14
Retentivity(M_r) (memu)	277.10	1593.30	692.13
Mass(M) (g) (10^{-3})	36	35	43

Pure NiO structure has an antiferromagnetic behavior and it also suggests that ferromagnetic behavior enhances by substituting the alkali metal ion in the host lattice. The observed NiO nanostructure shows weak ferromagnetic behavior with a magnetization of 6.3815 memu/g, since a modest amount of extra charge can switch the mutual alignment of the magnetization from antiferromagnetic to ferromagnetic. At x = 0.03 Li concentration, the cubic NiO attains the ferromagnetic behavior with high magnetization of 18.476 memu/g. The interlayer magnetic order can be reversed by absorption of alkali metal on magnetic layer. After high doping at x = 0.06, ferromagnetic behavior decreases due to excess alkali metal absorption in the interlayer of NiO. Hence, the ferromagnetic behavior loses its magnetization as 8.4295 memu/g.

Conclusions

The XRD results show that the cell constant of the prepared sample decreases with the concentration of Li-dopant. This ensures that Li occupies host lattice of Ni. The inter atomic distance decreases with the increase of Li content from x = 0.03 and x = 0.06 which is due to the higher ionic radius of Li^+ (r_{Li} = 0.76 Å) than Ni^{2+} (r_{Ni} = 0.69 Å). The electronic distributions in the unit cell has been analyzed through the MEM method for the prepared nanostructures and the bonding features are found to behave like covalent material. Higher peak intensity is observed for the XRD profile of $Ni_{0.97}Li_{0.03}O$. The UV analysis shows that the band gap decreases for $Ni_{0.97}Li_{0.03}O$ nanostructure and the MEM analysis shows an enhanced charge distribution along the bonding direction. Pure NiO structure has antiferromagnetic behavior and it also suggests that ferromagnetic behavior enhances by substituting the alkali metal ion in the host lattice. The observed NiO nanostructure shows weak ferromagnetic behavior. At x = 0.03 Li concentration, the cubic NiO attains the ferromagnetic behavior with high magnetization. For further higher doping at x = 0.06, ferromagnetic behavior decreases.

References

[1]   H. Sato, T. Minami, S. Takata, T. Yamada, Transparent conducting p -type NiO thin films prepared by magnetron sputtering, Thin Solid Films 23 (1993) 27-31. http://dx.doi.org/10.1016/0040-6090(93)90636-4

[2]   X. Luo, L.T. Tseng, S. Li, J.B. Yi, Room temperature ferromagnetic ordering of NiO films through exchange coupling, Mat. Sci. Semicon. Proc. 30 (2015) 228–232. http://dx.doi.org/10.1016/j.mssp.2014.10.009

[3]   S. Saravanakumar, R. Saravanan, S. Sasikumar, Effect of sintering temperature on magnetic properties and charge density distribution of nano-NiO, Chem. Pap. 68 (2014) 788–797. http://dx.doi.org/10.2478/s11696-013-0519-1

[4]   I. Sugiyama, N. Shibata, Z. Wang, S. Kobayashi, T. Yamamoto, Y. Ikuhara, Ferromagnetic dislocations in antiferromagnetic NiO, Nat. Nanotechnol. 8 (2013) 266–270. http://dx.doi.org/10.1038/nnano.2013.45

[5]   F.H. Aragón, P.E.N. de Souza, J.A.H. Coaquira, P. Hidalgo, D. Gouvêa, Spin-glass like behavior of uncompensated surface spins in NiO nanoparticulated powder, Physica B 407 (2012) 2601–2605. http://dx.doi.org/10.1016/j.physb.2012.04.003

[6]   R.M. Gabr, A.N. EI-Naimi, M.G. AI-Thani, Preparation of nanometer nickel oxide by the citrate-gel process. Thermochim. Acta 197 (1992) 307-318. http://dx.doi.org/10.1016/0040-6031(92)85029-U

[7]   M. Yoshio, Y. Todorov, K. Yamato, H. Noguchi, J. Itoh, M. Okada and T. Mouri, Synthesis and characterization of NiO nanoparticles by sol-gel method, Materials transactions 74 (1998) 46-53. http://dx.doi.org/10.4028/www.scientific.net/amr.123-125.181

[8]   S.C. Chen, T.Y. Kuo, Y.C. Lin, C.L. Chang, Preparation and properties of p-type transparent conductive NiO films, Adv Mater Res. 123 (2010) 181–184.

[9]   R.C. Korosec, P. Bukovec, Sol–gel prepared NiO thin films for electrochromic applications. Acta Chim Slov. 53 (2006) 136–147.

[10]   I.M. Chan, F.C. Hong, Improved performance of the single-layer and double-layerorganic light emitting diodes by nickel oxide coated indium tin oxide anode.Thin Solid Films 450 (2004) 304–311. http://dx.doi.org/10.1016/j.tsf.2003.10.022

[11]   I. Hotovy, J. Huran, P. Siciliano, S. Capone, L. Spiess, V. Rehacek, Enhancement of H2 sensing properties of NiO-based thin films with a Pt surface modification,

Sens Actuator B-Chem. 103 (2004) 300–311.
http://dx.doi.org/10.1016/j.snb.2004.04.109

[12]   A.Krier, M. Yin, V .Smirnov, et al., The development of room temperature LEDs
       and Lasers for mid-infrared spectral range, Phys Stat Sol A. 205 (2008)129-143.
       http://dx.doi.org/10.1002/pssa.200776833

[13]   H. Ohtah, M. Kamiya, T. Kamiya, M. Hiran, H. Hosono, UV-detector based on
       pn-heterojunction diode composed of transparent oxide semiconductors, p-NiO/n-
       ZnO,Thin Solid Films 445 (2003) 317–321. http://dx.doi.org/10.1016/S0040-
       6090(03)01178-7

[14]   A.E. Berkowitz and K. Takano, Exchange anisotropy - a review, J.Magn. Mater.
       200(1999) 552-570. http://dx.doi.org/10.1016/S0304-8853(99)00453-9

[15]   V. Skumryev, S. Stoyanov, Y. Zhang, G. Hadjipanayis, D. Givord and J. Nogue´,
       Ordered magnetic nanostructures, Nature 423 (2003) 850-853.
       http://dx.doi.org/10.1038/nature01687

[16]   K.C. Liu, M.A. Anderson, J. Porous nickel oxide/nickel films for electrochemical
       capacitors, Electrochem. Soc. 143 (1996) 124–130.
       http://dx.doi.org/10.1149/1.1836396

[17]   Y.D. Wang, C.L. Ma, X.D. Sun, H.D. Li, Preparation of nanocrystalline metal
       oxide powders with the surfactant-mediated method, Inorg. Chem. Commun. 5
       (2002) 751–755. http://dx.doi.org/10.1016/S1387-7003(02)00546-4

[18]   L. Xiang, X.Y. Deng, Y. Jin, Experimental study on synthesis of NiO nano-
       particles, Scripta. Mater. 47 (2002) 219–224. http://dx.doi.org/10.1016/S1359-
       6462(02)00108-2

[19]   E.R. Beach, K.R. Shqaue, S.E. Brown, S. J. Rozesveld, P.A. Morris, Solvothermal
       synthesis of crystalline nickel oxide nanoparticles, Mater. Chem. Phys. 115 (2009)
       373–379. http://dx.doi.org/10.1016/j.matchemphys.2008.12.018

[20]   S. Deki, H. Yanagimito, S. Hiraoka, NH2-terminated poly (ethylene oxide)
       containing nanosized NiO particles: synthesis, characterization, and structural
       considerations, Chem. Mater. 15 (2003) 4916–4922.
       http://dx.doi.org/10.1021/cm021754a

[21]   H.M. Rietveld, The Rietveld Method, J. Appl. Crystallogr. 2 (1969) 655-697.
       http://dx.doi.org/10.1107/S0021889869006558

[22]   V. Petricek, M. Dusek, L. Palatinus Jana, The crystallographic computing system
       Institute of Physics, Praha, Czech Republic, (2006).

[23] S. Kumazawa, Y. Kubota, M. Takata, M. Sakata, Y. Ishibashi, Electron-density-distribution Calculation by the maximum-entropy method, *Journal* of *Applied Crystallography*. 26 (1993) 453-457.
http://dx.doi.org/10.1107/S0021889892012883

[24] F. Izumi, R.A. Dilanian, Recent research developments in physics Part II, vol. 3 (Transworld Research Network, Trivandrum, (2002).

Keywords

About the Editor

Dr Ramachandran Saravanan, has been associated with the Department of Physics, The Madura College, affiliated with the Madurai Kamaraj University, Madurai, Tamil Nadu, India from the year 2000. He is the head of the Research Centre and PG department of Physics. He worked as a research associate during 1998 at the Institute of Materials Research, Tohoku University, Sendai, Japan and then as a visiting researcher at Centre for Interdisciplinary Research, Tohoku University, Sendai, Japan up to 2000.

Earlier, he was awarded the Senior Research Fellowship by CSIR, New Delhi, India, during Mar. 1991 - Feb.1993; awarded Research Associateship by CSIR, New Delhi, during 1994 – 1997. Then, he was awarded a Research Associateship again by CSIR, New Delhi, during 1997- 1998. Later he was awarded the Matsumae International Foundation Fellowship in1998 (Japan) for doing research at a Japanese Research Institute (not availed by him due to the simultaneous occurrence of other Japanese employment).

He has guided six Ph.D. scholars as of 2016, and about ten researchers are working under his guidance on various research topics in materials science, crystallography and condensed matter physics. He has published around 100 research articles in reputed Journals, mostly International, apart from around 45 presentations in conferences, seminars and symposia. He has also guided around 50 M.Phil. scholars and an equal number of PG students for their projects. He has attracted government funding in India, in the form of Research Projects. He has completed two CSIR (Council of Scientific and Industrial Research, Govt. of India), one UGC (University Grants Commission, India) and one DRDO (Defense Research and Development Organization, India) research projects successfully and is proposing various projects to Government funding agencies like CSIR, UGC and DST.

He has written 3 books in the form of research monographs with details as follows; "Experimental Charge Density - Semiconductors, oxides and fluorides" (ISBN-13: 978-3-8383-8816-8; ISBN-10:3-8383-8816-X), "Experimental Charge Density - Dilute Magnetic Semiconducting (DMS) materials" (ISBN-13: 978-3-8383-9666-8; ISBN-10: 3-8383-9666-9) and "Metal and Alloy Bonding - An Experimental Analysis" (ISBN -13: 978-1-4471-2203-6). He has committed to write several books in the near future.

His expertise includes various experimental activities in crystal growth, materials science, crystallographic, condensed matter physics techniques and tools as in slow evaporation, gel, high temperature melt growth, Bridgman methods, CZ Growth, high vacuum sealing etc. He and his group are familiar with various equipment such as: different types of cameras; Laue, oscillation, powder, precession cameras; Manual 4-

circle X-ray diffractometer, Rigaku 4-circle automatic single crystal diffractometer, AFC-5R and AFC-7R automatic single crystal diffractometers, CAD-4 automatic single crystal diffractometer, crystal pulling instruments, and other crystallographic, material science related instruments. He and his group have sound computational capabilities on different types of computers such as: IBM – PC, Cyber180/830A – Mainframe, SX-4 Supercomputing system – Mainframe. He is familiar with various kind of software related to crystallography and materials science. He has written many computer software programs himself as well. Around twenty of his programs (both DOS and GUI versions) have been included in the SINCRIS software database of the International Union of Crystallography.

www.ingramcontent.com/pod-product-compliance
Lightning Source LLC
Chambersburg PA
CBHW071650210326
41597CB00017B/2172